Inquiry-Based
Experiments in Chemistry

Inquiry-Based Experiments in Chemistry

Valerie Ludwig Lechtanski

AMERICAN CHEMICAL SOCIETY
Washington, D.C.

OXFORD
UNIVERSITY PRESS

2000

OXFORD
UNIVERSITY PRESS

Oxford New York
Athens Auckland Bangkok Bogotá Buenos Aires Calcutta
Cape Town Chennai Dar es Salaam Delhi Florence Hong Kong Istanbul
Karachi Kuala Lumpur Madrid Melbourne Mexico City Mumbai
Nairobi Paris São Paulo Singapore Taipei Tokyo Toronto Warsaw

and associated companies in
Berlin Ibadan

Developed and distributed in partnership by
American Chemical Society and Oxford University Press

Published by Oxford University Press, Inc.
198 Madison Avenue, New York, New York 10016

Library of Congress Cataloging-in-Publication Data
Lechtanski, Valerie Ludwig, 1958–
Inquiry-based experiments in chemistry / Valerie Ludwig Lechtanski
p. cm.
Includes bibliographical references and index.
ISBN 0-8412-3570-8
1. Chemistry—Laboratory manuals. I. Title.
QD45.144 1999 98-53217
540'78—dc21

Word of Caution:
This volume is intended for use by professional chemists and chemistry teachers. Even the simplest chemistry activity is potentially dangerous when performed by someone lacking the manipulative skills and knowledge of chemistry necessary to understand the reactions involved. Every precaution must be taken to ensure the safety of the instructor and the students. You should follow the directions given for each activity and not use materials that aren't recommended.

1 3 5 7 9 8 6 4 2

Printed in the United States of America
on acid-free paper

This book is dedicated to my entire family

for their unwavering support and encouragement.

Preface

When people know how scientists go about their work and reach scientific conclusions, and what the limitations of those conclusions are, they are more likely to react thoughtfully to scientific claims and less likely to reject them out of hand or accept them uncritically.

One of the goals of teaching science should be teaching the *process* of science.[1] Unfortunately, the typical secondary chemistry lab manual includes experiments that provide a detailed procedure, directions regarding what data to obtain and when to take particular measurements, organized data tables, and the step-by-step calculations or analysis required to reach a conclusion. These pre-fab or "cookbook" labs do not give students the opportunity to learn or appreciate the process of science.

The 1996 National Science Education Standards (NSES) expect teachers to plan and incorporate inquiry into the science curriculum. Some of the student outcomes listed in the NSES document include the ability to design and conduct scientific investigations, formulate scientific explanations using experimental evidence, and effectively communicate the results of scientific investigations.[2]

Research has shown that students using a laboratory-investigative approach show significant gains in formulating hypotheses, making assumptions, designing and executing investigations, understanding variables, making careful observations, recording data, analyzing and interpreting results, and synthesizing new knowledge, as well as the development of curiosity, openness, responsibility, and satisfaction. These students demonstrate a "better ability to think scientifically."[3]

The question then becomes, How can teachers successfully incorporate inquiry into the laboratory without overwhelming themselves or the students? Some suggestions include:[4]

- choose labs that address simple concepts
- choose experiments that can be completed using familiar equipment
- choose a lab where the data can be pooled by the class and that lends itself to the determination of a mathematical relationship
- reduce the procedural steps of the experiment
- increase the opportunities for the students to think about the data they should collect.

The experiments chosen for this book attempt to accomplish these goals. They use an inquiry approach, involve straightfoward concepts, use familiar equipment, require students to design some or all of the procedure, and provide opportunities for them to think about the data they should collect as well as its analysis.

A concern often addressed by secondary teachers when discussing the use of the inquiry method is the extended amount of lab time that they assume is required. Instructors are often concerned that introducing the inquiry method will be at the expense of content. The labs in this book intentionally involve concepts that are straightforward and therefore should not significantly increase the required lab time or impact the curricular content of the course. I hope that by providing basic labs in an inquiry context teachers will be more likely to use them in their classrooms, thereby providing an introduction to the scientific process. (Ideally, students will also be given an opportunity during the year—such as with a science fair project—to perform more in-depth experimentation).

Clearly, not all labs performed in the secondary classroom should be inquiry based. When performing experiments involving serious safety hazards, students should be provided with clear instructions and detailed procedures. However, a mixture of traditional labs and inquiry-based labs will provide students with a strong chemistry background as well as a better understanding of the scientific process.

TARGET AUDIENCE

The experiments included in this book are intended to be used by first-year chemistry students or students enrolled in an introductory physical science course. Students accustomed to the "cookbook"-style lab typically experience some initial frustration when asked to design their own procedure but after performing a few labs will become more independent and begin to trust their own ideas.

THE ROLE OF THE TEACHER

In inquiry learning, the teacher does not abandon the students but acts as a facilitator, assisting and guiding students as needed. The teacher might ask questions that force the students to think about what they are doing, such as: What do you think will happen? Would the sample size impact the outcome? Do you need to repeat any steps or measurements? Is there another way to perform the same investigation? What are some sources of error and could they be minimized if the experiment was repeated?

CHOOSING EXPERIMENTS FOR THE BOOK

Each of the experiments included in this book has been run many times by first-year chemistry students. The labs included here are intended to address concepts commonly covered in a first-year chemistry course, can be designed by the students with minimal risks, and can be completed in one or two lab periods.

FORMAT OF THE BOOK

Each experiment is divided into three parts:

the experiment (written for the students)
teacher's notes
sample lab report.

The Experiment

This page is directed to the students who will be performing the lab. It includes the purpose of the experiment, background information on the topic to be studied, safety information, and questions for further thought. In a few cases, students are also given a basic outline of the procedure if safety hazards are associated with the experiment or if the lab is intended to introduce a particular technique. In these cases, students are still given the task of deciding some parameter(s) of the general procedure.

The Teacher's Notes

Each lab includes teacher's notes intended to give you further information regarding the execution of the experiment. The teacher's notes include:

Options: The purpose of this section is to provide you with some extensions or options when running the lab. These options may be used to adjust the experiment to the particular level of the class, provide a follow-up experiment, or allow opportunities for student creativity.

Notes: This section includes where materials can be purchased, helpful hints, and other background information that you may find useful.

Time: The times given in the teacher's notes represent the average time needed by first-year chemistry students. As noted in this section, the time needed for students to design and execute their own experiment can vary depending on the level and creativity of the class, their familiarity with designing their own experiments, and the complexity of the procedure they develop. The time needed to obtain teacher approval is not included in this section, as some teachers will accept a verbal description of the planned experiment while other teachers want a detailed written procedure. You should adjust the times accordingly.

Teams: A recommended number of students per team is provided. Although some experiments can be performed individually, working in teams stimulates a greater exchange of ideas and provides students with the opportunity to work cooperatively. Typically, the more involved the experiment (so that students may have difficulty deciding on a procedure), the more students recommended per team.

Materials: This section lists the materials commonly used by students when executing the experiment. It is recommended that you have the materials available, but not display them to the students ahead of time. Seeing the supplies before designing their procedure can cause students to limit their experiment to one that only uses those materials, or may cause students to think that the only "right" procedure involves the given equipment.

Sample Procedures: The sample procedures given are those typically designed by first-year students. This section cannot anticipate every experiment that students may design but should provide you with a general idea of what to expect.

Safety Precautions: You should refer to the safety precautions given in the experiment as well as in the teacher's notes. The safety precautions are intended to point out hazards associated with the chemicals and experiment. You should always require the use of goggles, gloves, and lab coats (or aprons) and have Material Safety Data Sheets (MSDS) for each substance used available to the students.

Common Misconceptions, Procedural, and Calculation Errors: Students typically make many of the same errors and assumptions when designing experiments. This section indicates some of the more common errors associated with the lab. By reading this section, you can help students avoid some common pitfalls.

Lab Report: If students will be writing a formal lab report, this section of the teacher's notes indicates which aspects of the experiment could be included in the report.

Answers to the Questions: Each experiment includes additional questions that can be answered by the students. This section contains the answers to those questions.

For Further Reading: Additional information regarding the topic of the lab can be found in the references cited in this section.

Sample Lab Report

The sample lab report given for each activity is indicative of a report commonly written by first-year students and demonstrates the most common procedure used by the students. It also provides data that are representative of that usually obtained by the students. The sample lab report lists a composite of the sources of error that students typically include in their lab report.

GETTING STARTED

Before beginning any of the labs in the book, students should be familiar with:

1. all safety information and procedures (see appendix 1),
2. the location and use of safety equipment,
3. the location of MSDS sheets,
4. how to properly use a balance (electronic or other),
5. how to safely light a Bunsen burner (unless hot plates will be used), and
6. the equipment and supplies that are available. You may want to show, and demonstrate how to safely use, the following equipment:

   goggles, gloves, and lab coat/apron
   beakers
   buret
   dropper pipet
   Erlenmeyer flasks
   funnel

gas-measuring tube
gas-collecting (or wide mouth) bottle
graduated cylinder
stirring rod
test tubes
test tube rack
thermometer
watch glass
evaporating dish
Bunsen burner
ring stand
iron ring
wire gauze
tongs
forceps
test tube clamp
pneumatic trough
mortar and pestle
pinch clamp
test tube holder
weighing paper or weighing dishes

APPROVING THE PROCEDURE

Students are directed in the lab to obtain the approval of their instructor before beginning the experiment. When approving a procedure:

A. Ensure that the procedure does not involve any possible safety risks.

B. Specify to the students the form in which the procedure should be presented for approval. For example, some instructors will accept a verbal description of the procedure such as: "We are going to take three 1.0 g samples of the crystals and add each sample to 100 mL of water in a 400 mL beaker. The water in one beaker will be at room temperature, one will be at 40 °C, and one will be at 60 °C. Then we will time how long it takes for the crystals to dissolve in each beaker", while other instructors want a detailed written procedure. (You can also vary the requirement from experiment to experiment.) If you require a written procedure, you may want to keep a copy so students can't make changes in their experimental design without permission.

C. Decide if you will only approve a procedure that is "right", that is, it will provide meaningful data to the students. Some instructors are comfortable in allowing students to make and correct their mistakes, whereas other instructors will not approve a procedure until potential errors have been removed.

D. Determine if you will allow the students the freedom to go back and make adjustments to a procedure if they find an error after beginning the lab, for example, if something occurs that they hadn't anticipated or if the experiment isn't working like they had expected. This can be a good learning experience for the students, but you should make it clear that they cannot change a procedure without receiving approval, and you should watch for students who are merely changing their procedure because another team's idea looks better.

E. Consider withholding chemicals and other materials until the procedure has been approved.

THE LAB REPORT

For experiments that you want students to write a formal lab report, each individual on the team can write a report, or you can choose to have one report per team. If each team writes a single report, you may (1) have each person on the team responsible for some aspect of the report. For example, one person may be in charge of recording the procedure and data, while the other person performs the calculations and summarizes the sources of error. (You may want to stipulate how the lab report should be divided among team members and require that each member sign their name to their contribution) or (2) have the team members take turns being responsible for the lab write-up from experiment to experiment (i.e., students can take turns being the team captain, although each team member should still sign their name indicating that they agree with the information included).

Whenever students are writing a team report, the instructor should make it clear that the report must represent everyone's work and ideas. All team members should check the calculations, conclusion, sources of error, and so on before handing in the report.

The aspects of a lab report that should be included for a given experiment are listed in the associated teacher's notes and explained below.

Title: this may be the same title as given in the experiment.
Purpose: this may be copied from the lab.
Materials: a list of materials used in the experiment.
Procedure: a numbered list of steps indicating the procedure followed in the experiment.
Data: data should be neatly organized into tables or charts.
Calculations: when performing calculations students should show all work, label the calculations (indicate what is being calculated), include units in the work and on the answer, and round the answer to the correct number of significant figures.
Graphs: a graph is only required in certain experiments.
Conclusion and Sources of Error: in this section, students should discuss all of the potential sources of error identified during the experiment and state a final conclusion.
Percent Error: students may be asked to calculate the percent error for some experiments.

ALTERNATIVES TO WRITING A LAB REPORT

Reading and grading a lab report for every experiment and student can be cumbersome and time consuming. For some experiments, instructors may decide to use one of the following alternatives:

Compile a set of class data and compare results,
Have each team of students perform one aspect of the lab and share the results with the other members of the class,
Use the data to learn how to calculate a particular value (such as the heat of fusion of ice) and then evaluate the students' understanding on a written exam,
Simply record the results in their notebooks, or
Make an oral presentation to the class describing their procedure and results.

ASSESSING THE LAB REPORT

You may want to use some of the following criteria when evaluating written lab reports:

1. Is the purpose of the experiment clearly stated?
2. Does the procedure directly address the purpose?
3. Is the procedure written in such a way that it could be easily repeated by another person?
4. Were the measurements recorded correctly?
5. Are data presented clearly and neatly?
6. Are the data meaningful and reliable?
7. Are the calculations organized clearly?
8. Are the calculations done correctly, including the use of units and significant figures?
9. Does the conclusion address the purpose?
10. Does the conclusion relate to the experimental results?
11. Did everyone on the team contribute?
12. Did the student(s) demonstrate proper laboratory techniques?
13. Did the students clean the lab area at the conclusion of the lab?
14. Were all safety rules observed?

REFERENCES

1. Benchmarks for Science Literacy: Project 2061. American Association for the Advancement of Science. Oxford University Press, New York. 1993. p. 3.
2. Chemistry in the National Science Education Standards. American Chemical Society, Washington, DC, 1997. p. 19.
3. Raghubir, K.P. *Journal of Research in Science Teaching.* 1979, pp. 13–17.
4. *Chemistry in the National Science Education Standards.* American Chemical Society, Washington, DC, 1997. pp. 22–23.

Acknowledgement

I would like to acknowledge Laura Gigliotti and Paul E. Ludwig for their help with some of the graphics in the book. You are the best brother and sister anyone could have!

Contents

Inquiry-Based
Experiments in Chemistry

1 Identifying Unknown Metals

THE EXPERIMENT

PURPOSE

To determine the identity of unknown metals by calculating their densities.

BACKGROUND

The density (D) of a substance is the ratio of its mass (M) to its volume (V).

The formula for density is $D = M / V$.

You will be given samples of two or more metals and asked to identify the metals by calculating their densities and comparing the experimental values to a list of accepted values.

PROCEDURE

NOTE: Consult safety information and obtain teacher approval before beginning the experiment.

Design an experiment to determine the density and identity of the unknown metals as accurately as possible.

SAFETY INFORMATION

After reviewing your procedure, the instructor will discuss any safety precautions that are specific to your experiment.

TABLE OF ACCEPTED VALUES

METAL	DENSITY (g/cm^3)
Aluminum	2.7
Zirconium	6.4
Zinc	7.14
Chromium	7.19
Tin	7.31
Iron	7.86
Nickel	8.9

METAL	DENSITY (g/cm^3)
Copper	8.92
Lead	11.34
Platinum	21.45

QUESTIONS FOR FURTHER THOUGHT

1. Is density an intensive or extensive property? Is density a chemical or physical property?
2. Are the mass and volume of a given substance directly or inversely proportional?
3. Sketch a graph of mass versus volume for a given substance.
4. How could density be used to determine the concentration of antifreeze in the cooling system of an automobile?

1 Identifying Unknown Metals

TEACHER'S NOTES

The purpose of this introductory experiment is more than simply having the students calculate the density of unknown metals. Another objective of this lab is to introduce some of the characteristics of a good experiment, such as performing multiple trials, organizing and labeling data and calculations correctly, identifying sources of error, recording measurements correctly, the proper use of significant figures in calculations, and writing an organized lab report. The topic of density has purposefully been selected as an introductory lab because it is undoubtedly familiar to high school chemistry students and therefore allows students to focus on the process of experimentation.

You can choose to have the students perform the experiment either before or after discussing the components of a good experiment. Either way, this experiment (and the other experiments involving density) can lead to discussions that set the framework for the remaining experiments in this book.

OPTIONS

1. The first four labs in the book involve the density of metals. The intention is not that each student will perform all of them. Each of the labs introduces a different aspect of density and reinforces the concepts involved in designing an experiment. You may: (a) choose one density experiment for the entire class, or (b) have each team of students perform a different experiment involving density and then report the results to the class. The second option works well when first trying to develop the aspects of a good experiment. Showing the results of the various labs can lead to a discussion of the different methods used, various sources of error, recording measurements correctly, using significant figures in calculations, and so on.

2. If individual teams will be performing different labs on density and you need more experiments to provide each team with a unique activity, other ideas include: (a) comparing the density of several liquids (water, isopropyl alcohol, vegetable oil, corn syrup, glycerin, etc.), or (b) comparing the density of pennies made before 1982 and after 1982. (The composition of pennies changed in 1982 from pure copper to copper over a zinc core.)

NOTES

1. Encourage the students before they begin to design the best experiment possible. (Perhaps there could be a reward for the team that has the least amount of error in their final density values.) Explain to them that one of the goals of the lab is to develop an understanding of what constitutes a good experiment and that this is one of the aspects you will be evaluating.

2. Have beakers of the unknown metals prepared and labeled for the students. The metals can be identified as A, B, C, and so on. Have a sample of the metal available to each group that is large enough for students to perform more than one trial without having to reuse any wet metal shot, (although this may lead students to believe that they have to use the entire

sample in their experiment). You may want to mention that the students can use all or part of the sample.

3. If the students are not familiar with using water displacement to measure the volume of an irregularly shaped object, you can discuss this with the students ahead of time.

4. To avoid having students accidentally pour metal BBs into the sink, you may want to have the students pour the metal shot and water (after measuring the volume) into a funnel with filter paper and have a container available for the wet metals. This will prevent the wet BBs from being added back to the original container of dry BBs, which can affect the data of an incoming class. A plastic shoe box lined with paper towels allows the metal shot to dry quickly and be ready for another lab session.

5. Students may initially be uncomfortable with designing their own experiment and seek constant approval and assistance. You should be supportive and encouraging. After performing several experiments in which they have designed their own procedure, the students will become more confident and independent.

TIME

The amount of time needed for students to design and execute their own experiment can vary depending on the level and creativity of the class, their familiarity with designing their own investigations, and the complexity of the procedure they develop.

Average Time for This Experiment

Time needed to design the experiment: 15 minutes (if students are familiar with the technique of water displacement)
Time needed to run the experiment: 30 minutes

TEAMS

Teams of two work well. Although students can perform this lab individually, working in groups stimulates a greater discussion of the techniques and procedures to be used.

MATERIALS (PER TEAM)

Require the use of goggles, gloves, and aprons/lab coats. The following supplies should be readily available:

- Metals (3 or 4 of the following metals):
  - Zinc (strips, mossy zinc, 1″ rectangles, etc.)
  - Copper (shot, wire, strips, etc.)
  - Aluminum (shot, wire, etc.)
  - Lead shot (see safety precaution)
  - Iron nails
- 50-mL or 100-mL graduated cylinder
- Balance
- Weighing paper or weighing dishes
- Funnel and filter paper (if students will be draining the metals)
- Beakers (any size)

NOTE: The densities of other metals are given in the table of accepted values, but they are not recommended for use (see safety precautions below).

SAMPLE PROCEDURES

Most students will think this is an easy experiment and start right away. Typically, they will measure the mass of the metal and then use the technique of water displacement to measure its volume.

However, they may not consider the following:

1. Doing several trials with each metal to minimize error,
2. Trying more than one method to determine the volume of the metal such as $V = L \times W \times H$ if using rectangular pieces of metal, or
3. Discussing what sample size to use or how the measurements for mass and volume will affect the accuracy of the density (in terms of significant figures). For example, using a sample that has a volume of 0.6 mL will limit the resulting density value to one significant figure, making it more difficult to identify the unknown metal.

SAFETY PRECAUTIONS

Discuss with the students any safety precautions that are specific to the procedure they develop. Material Safety Data Sheets for each substance used should be reviewed and made available to the students.

The use of metal powders is not recommended, as many fine powders can form explosive mixtures in the air.

The density of chromium is given in the table of accepted values, but its use is not recommended because it is a known carcinogen.

Lead shot is an optional metal in this lab, although it can be fatal if swallowed. Take appropriate precautions.

COMMON MISCONCEPTIONS, PROCEDURAL AND CALCULATIONS ERRORS

1. Students might measure the volume of the metal before measuring the mass. This will affect the mass measurement since the metal will be wet.

2. Students might think that they have to use the same number of BBs (or the same mass of BBs) in each trial. This would demonstrate that the students do not understand the proportional relationship between mass and volume for a given substance.

3. Students might use one BB for each trial. The volume change for this will be so small for some forms of metal shot that it will be difficult to measure.

4. Students may determine the volume of the BBs by pouring a sample of the dry metals into a beaker or cylinder, that is, they don't use water displacement. This doesn't account for the volume occupied by the air found in between the BBs.

5. Students may not realize that 1 mL = 1 cm^3 and therefore will not know how to compare their calculated density to those given in the table of accepted values.

6. Students may determine the volume of the metal by performing water displacement in a beaker rather than in a graduated cylinder.
7. Students might only perform one trial with each metal.

LAB REPORT

There are several alternatives to writing a formal lab report given in the introduction of the book. If students are writing a formal lab report, it should include the following:

I. Title
II. Purpose
III. Materials
IV. Procedure
V. Data
VI. Calculations
VII. Conclusion (including a discussion of sources of error)
VIII. Answers to the Questions (optional)

ANSWERS TO THE QUESTIONS

1. Density is an intensive property as well as a physical property.
2. The mass and volume of a given substance are directly proportional.
3. A graph of mass versus volume would be a straight line that passes through the origin.
4. Because the density of water and ethylene glycol (antifreeze) are different, the density of the solution in the cooling system can be used to determine the concentration of the mixture and therefore how well it will protect against freezing. A hydrometer is used to determine the density of the antifreeze solution in the radiator.

FOR FURTHER READING

Wieczorek, L. *The Science Teacher.* May 1995, 62, pp. 44–45.

1 Identifying Unknown Metals

SAMPLE LAB REPORT

PURPOSE

To determine the identity of four unknown metals by calculating their densities.

MATERIALS

Safety equipment: goggles, apron, gloves
4 Unknown metals
Balance

100-mL graduated cylinder
Funnel
Filter paper
Weighing dishes

PROCEDURE

NOTE: The student is only performing one trial with each metal.

1. Find the mass of an empty weighing dish.
2. Add approximately 30.00 g of metal A to the dish. Record the mass of the weighing dish and metal.
3. Pour 50.0 mL of tap water into the 100-mL graduated cylinder.
4. Add metal A to the cylinder.
5. Record the volume of the water and metal A.
6. Pour the metal and water into a funnel with filter paper.
7. Return the wet metal to the instructor.
8. Repeat steps 1–6 with metals B, C, and D.

DATA

METAL	A	B	C	D
1. Mass of empty weighing dish	0.97 g	1.01 g	0.99 g	0.99 g
2. Mass of dish and metal	34.32 g	36.83 g	34.43 g	34.61 g
3. Final volume of water in cylinder	62.5 mL	54.2 mL	54.1 mL	54.9 mL

CALCULATIONS

METAL	A	B	C	D
Mass of metal (Measurement #2 − #1)	33.35 g	35.82 g	33.44 g	33.62 g
Volume of metal (Measurement #3 − 50.0 mL)	12.5 mL	4.2 mL	4.1 mL	4.9 mL
Density of metal = m/v	2.67 g/mL	8.5 g/mL	8.2 g/mL	6.9 g/mL

CONCLUSION

Based on the values that we calculated from our data and comparing those values to the table of accepted values, we believe the metals to be:

A: aluminum
B: copper (the density value was also close to nickel, but metal B had the color of copper metal)
C: iron
D: zinc

SOURCES OF ERROR

The following is a compilation of sources of error that students may suggest:

1. The metals may not be pure.
2. The metals may be wet from another class.

3. Some of the metals look like they were corroded.
4. Water may have splashed out of the cylinder when we added the metal.
5. Air may have been trapped in between the BBs, which would alter the volume measurement.
6. Human error and the accuracy of the instruments.

2 Is Density an Intensive or Extensive Property of Matter?

THE EXPERIMENT

PURPOSE

To determine if density is an intensive or extensive property of matter.

BACKGROUND

Intensive properties of a substance are properties that *do not change* when the mass of the sample is varied. Extensive properties of a substance are properties that *do change* when the mass of the sample is varied. You will be given a sample of a metal and asked to design an experiment to determine if density is an intensive or extensive property of matter.

PROCEDURE

NOTE: Consult safety information and obtain teacher approval before beginning the experiment.

Design an experiment to determine if density is an intensive or extensive property of matter.

SAFETY INFORMATION

After reviewing your procedure, the instructor will discuss any safety precautions that are specific to your experiment.

QUESTIONS FOR FURTHER THOUGHT

1. Identify each of the following properties as an intensive property or an extensive property of matter:
   a. boiling point
   b. freezing point
   c. volume
   d. mass
   e. length
2. If you had an unknown substance and were asked to identify it, would you use the intensive properties of the substance or the extensive properties of the substance? Why?

2 Is Density an Intensive or Extensive Property of Matter?

TEACHER'S NOTES

Students will often know the answer to the question posed in this lab, but the purpose of the initial labs on density is to cover the concept involved and teach the students the processes involved in designing their own experiments. This lab (as well as the other density labs) is intended to help students become familiar with the characteristics of a good experiment: taking measurements correctly, the correct use of significant figures in performing calculations, attempting to reduce potential sources of error, the components of a lab report, and so on.

OPTIONS

1. There are several labs on density included in the book and the intention is not that each student will perform all of them. Each of the labs introduces a different aspect of density and reinforces the concepts involved in designing an experiment. You may: (a) choose one density experiment for the entire class, or (b) have each team perform a different experiment and then report the results to the class. The second option works well when first trying to develop the aspects of a good experiment. Showing the results of the various labs can lead to a discussion of the different methods used, various sources of error, recording measurements correctly, using significant figures in calculations, and so on, thereby setting a good foundation for the rest of the experiments.

2. If individual teams will be performing different labs on density and you need more experiments to provide each team with a unique activity, other ideas include: (a) comparing the density of several liquids (water, isopropyl alcohol, vegetable oil, corn syrup, glycerin, etc.), or (b) comparing the density of pennies made before 1982 and after 1982. (The composition of pennies changed in 1982 from pure copper to copper over a zinc core.)

NOTES

1. Determine if you want to reveal the identity of the metal to the students. Be sure to have a sample large enough for the students to perform several trials without having to reuse wet metal.

2. If the students are not familiar with using water displacement to measure the volume of an irregularly shaped object, you can discuss this with the students ahead of time.

3. To avoid having students accidentally pour metal BBs into the sink, you may want to have the students pour the metal shot and water (after measuring the volume) into a funnel with filter paper and have a container available for the wet metals. This will prevent the wet BBs from being added to the original container of dry BBs, which can affect the data of an incoming class. A plastic shoe box lined with paper towels allows the metal shot to dry quickly and be ready for another lab session.

TIME

The amount of time needed for students to design and execute their own experiment can vary depending on the level and creativity of the class, their familiarity with designing their own investigations, and the complexity of the procedure they develop.

Average Time for This Experiment

Time needed to design the experiment: 15–20 minutes
Time needed to run the experiment: 30 minutes

TEAMS

Teams of two work well.

MATERIALS (PER TEAM)

Require the use of goggles, gloves, and aprons/lab coats. The following supplies should be readily available for the students.

- Metals:
  - Copper shot
  - Lead shot
  - Aluminum shot
  - Zinc pieces (1″ long rectangles)
  - Iron nails
- 50-mL or 100-mL graduated cylinder
- Balance
- Weighing paper or weighing dishes
- Funnel and filter paper (if students will be draining the metals)
- Beakers (any size)

SAMPLE PROCEDURES

Since the purpose of the experiment is to determine if the density of a metal varies with the mass of the sample, students will probably vary the mass of the samples and calculate the density of each sample. However, there can be a lot of variety in the number of trials performed and sample size. The class could discuss the impact of each of these variables on the final density value.

SAFETY PRECAUTIONS

Discuss with the students any safety precautions that are specific to the procedure they develop. Material Safety Data Sheets for each substance used should be reviewed and made available to the students.

The use of metal powders is not recommended, as many fine metal powders can form an explosive mixture with air.

COMMON MISCONCEPTIONS, PROCEDURAL AND CALCULATIONS ERRORS

1. Students might measure the volume of the metal before measuring the mass. This will affect the mass measurement as the metal will be wet.
2. Students might think that they have to count the number of BBs used in each trial.

3. Students might only perform one trial.
4. Students might use a beaker, rather than a graduated cylinder, to measure the volume of the metal sample.
5. Students might only use one BB for each trial, making the volume difficult to measure.

LAB REPORT

There are several alternatives to writing a formal lab report given in the introduction of the book. If students are writing a formal lab report, it should include the following:

I. Title
II. Purpose
III. Materials
IV. Procedure
V. Data
VI. Calculations
VII. Conclusion (including a discussion of sources of error)
VIII. Answers to Questions (optional)

ANSWERS TO THE QUESTIONS

1. a, intensive; b, intensive; c, extensive; d, extensive; e, extensive.
2. When identifying an unknown substance, intensive properties are used because they will not vary with different size samples.

FOR FURTHER READING

Wieczorek, L. *The Science Teacher.* May 1995, 62, pp. 44–45.

2 Is Density an Intensive or Extensive Property of Matter?

SAMPLE LAB REPORT

PURPOSE

To determine whether density is an intensive or extensive property of matter.

MATERIALS

Safety equipment: goggles, apron, gloves
Sample of a metal shot
100-mL graduated cylinder
Balance
Funnel
Filter paper
Weighing dishes

PROCEDURE

1. Divide the metal shot into three different sized samples.
2. Measure the mass of three weighing dishes.
3. Add one of the samples to a weighing dish and record the mass.
4. Add sample #2 to another weighing dish and record the mass.
5. Add sample #3 to another weighing dish and record the mass.
6. Pour 30.0 mL of water into the graduated cylinder.
7. Add sample #1 to the graduated cylinder and record the final volume.
8. Pour the metal and water into a funnel with filter paper for draining.
9. Repeat steps 6–8 with the other two samples of the metal.

DATA

SAMPLE	MASS OF EMPTY DISH	MASS OF DISH AND METAL SHOT	VOL. OF WATER AND METAL SHOT
1	1.01 g	4.72 g	31.2 mL
2	0.98 g	7.39 g	32.1 mL
3	0.97 g	11.18 g	33.4 mL

CALCULATIONS

SAMPLE	MASS OF METAL	VOLUME OF METAL	DENSITY OF METAL ($D = M/V$)
1	3.71 g	1.2 mL	3.1 g/mL
2	6.41 g	2.1 mL	3.1 g/mL
3	10.21 g	3.4 mL	3.0 g/mL

CONCLUSION

The density values obtained for the three trials were similar. Allowing for various sources of error, we conclude that density is an intensive property of matter.

SOURCES OF ERROR

The following is a compilation of sources of error that students may suggest:

1. Human and mechanical error.
2. The metal samples may not have the same degree of purity.
3. The metals may be wet from another class.
4. Some of the metals look like they have corroded.
5. Water may have splashed out of the cylinder when we added the BBs.

3 Does Density Vary with the Form of a Substance?

THE EXPERIMENT

PURPOSE

To determine if density varies with the form of a substance.

BACKGROUND

You will be given samples of a metal (such as copper) in various forms (e.g., shot, wire, strips). Determine if the density of the metal varies with the form.

PROCEDURE

NOTE: Consult safety information and obtain teacher approval before beginning the experiment.

Design an experiment to determine if density varies with the form of a substance.

SAFETY INFORMATION

After reviewing your procedure, the instructor will discuss any safety precautions that are specific to your experiment.

QUESTIONS FOR FURTHER THOUGHT

1. Compare the density of a substance in its solid phase, liquid phase, and gas phase. Are there any exceptions to this trend?
2. A method used to calculate a person's percent body fat is to submerge them in a tank of water. How does this technique work?

3 Does Density Vary with the Form of a Substance?

TEACHER'S NOTES

Students will often know the answer to the question posed in this lab, but the purpose of the initial labs on density is to cover the concept involved and teach the students the processes involved in designing their own experiments. The lab is also intended to help students become familiar with the characteristics of a good experiment: taking measurements correctly, the correct use of significant figures in performing calculations, attempting to reduce potential sources of error, the components of a lab report, and so on.

OPTIONS

1. There are several labs on density included in the book and the intention is not that each student will perform all of them. Each of the labs introduces a different aspect of density and reinforces the concepts involved in designing an experiment. You may: (a) choose one density experiment for the entire class, or (b) have each team of students perform a different experiment and then report the results to the class. The second option works well when first trying to develop the aspects of a good experiment. Showing the procedures and results of the various labs can lead to a discussion of the different methods used, various sources of error, recording measurements correctly, using significant figures in calculations, and so on, thereby setting a good foundation for the rest of the experiments.

2. If individual teams will be performing different labs on density and you need more experiments to provide each team with a unique activity, other ideas include: (a) comparing the density of several liquids (water, isopropyl alcohol, vegetable oil, corn syrup, glycerin, etc.), or (b) comparing the density of pennies made before 1982 and after 1982. (The composition of pennies changed in 1982 from pure copper to copper over a zinc core.)

NOTES

1. Determine if you want to reveal the identity of the metal to the students. If possible, have samples large enough for students to perform several trials.

2. If the students are not familiar with the technique of using water displacement to measure the volume of an irregularly shaped object, you can discuss this with the students ahead of time.

3. To avoid having students accidentally pour metal BBs into the sink, you may want to have the students pour the metal shot and water (after measuring the volume) into a funnel with filter paper and have a container available for the wet metals. This will prevent the wet BBs from being added back to the original container of dry BBs, which can affect the data of an incoming class. A plastic shoe box lined with paper towels allows the metal shot to dry quickly and be ready for another lab session.

TIME

The amount of time needed for students to design and execute their own experiment can vary depending on the level and creativity of the class, their familiarity with designing their own investigations, and the complexity of the procedure they develop.

Average Time for This Experiment

Time needed to design the experiment: 15–20 minutes
Time needed to run the experiment: 30 minutes

TEAMS

Teams of two work well.

MATERIALS (PER TEAM)

Require the use of goggles, gloves, and aprons/lab coats. The following supplies should be readily available for the students:

- Various forms of the sample metal, such as:
  - Zinc shot, zinc strips, zinc pieces, mossy zinc, zinc cylinder
  - Copper shot, copper strips, copper wire, copper tubing, copper cylinder
  - Aluminum shot, aluminum foil, aluminum strips, aluminum wire, aluminum cylinder
- 50-mL or 100-mL graduated cylinder
- Balance
- Weighing paper or weighing dishes
- Funnel and filter paper (if students will be draining the metals)
- Beakers (any size)

SAMPLE PROCEDURES

Since the purpose of the experiment is to determine if the density of a metal varies with form, students will probably design an experiment in which they measure the mass and volume of the various forms of the metal and calculate the density. Students may or may not perform more than one trial on each form.

SAFETY PRECAUTIONS

Discuss with the students any safety precautions that are specific to the procedure they develop. Material Safety Data Sheets for each substance used should be reviewed and made available to the students.

The use of metal powders is not recommended, as many fine metal powders can form an explosive mixture with air.

COMMON MISCONCEPTIONS, PROCEDURAL AND CALCULATIONS ERRORS

1. Students may measure the volume of the metal before measuring the mass. This will affect the mass measurement as the metal will be wet.

2. Students may think that they have to use the same mass of each form of the metal.
3. Students may only perform one trial.
4. Students may use a beaker, rather than a graduated cylinder, to measure the volume of the metal sample.

LAB REPORT

There are several alternatives to writing a formal lab report given in the introduction of the book. If students are writing a formal lab report, it should include the following:

I. Title
II. Purpose
III. Materials
IV. Procedure
V. Data
VI. Calculations
VII. Conclusion (including a discussion of sources of error)
VIII. Answers to Questions (optional)

ANSWERS TO THE QUESTIONS

1. For most substances the density is greatest in the solid phase and least in the gaseous phase. Water is an exception to this rule. Water is most dense in the liquid phase (4°C) rather than in the solid phase.
2. When a person is submerged in water, their volume can be measured by determining the volume of water displaced in the tank. Because fat and muscle have different densities, the mass and volume of the person can be used to determine the percentage of fat in the body.

FOR FURTHER READING

Wieczorek, L. *The Science Teacher.* May 1995, 61, pp. 44–45.

3 Does Density Vary with the Form of a Substance?

SAMPLE LAB REPORT

PURPOSE

To determine whether density varies with the form of a substance.

MATERIALS

Safety equipment: goggles, apron, gloves
Aluminum shot
Aluminum strip

Aluminum foil
Aluminum rod
Electronic balance
50-mL graduated cylinder
weighing dishes

PROCEDURE

1. Determine the mass of a sample of the aluminum shot.
2. Pour 30.0 mL of tap water into the graduated cylinder.
3. Add the shot to the cylinder, recording the final volume.
4. Repeat steps 1–3 with the other forms of aluminum.

DATA

MATERIAL	MASS (g)	FINAL VOLUME (mL)
Aluminum shot	14.93	35.7
Aluminum strip	1.34	30.8
Aluminum rod	17.61	36.8
Aluminum foil	5.17	32.6

CALCULATIONS

MATERIAL	VOLUME OF MATERIAL (FINAL VOL. – 30.0 mL)	DENSITY ($D = M/V$)
Aluminum shot	5.7 mL	2.6 g/mL
Aluminum strip	0.8 mL	2 g/mL
Aluminum rod	6.8 mL	2.6 g/mL
Aluminum foil	2.6 mL	2.0 g/mL

CONCLUSION

The density of all four samples of aluminum seems to be about the same. The variation in the final values could be due to the various sources of error.

SOURCES OF ERROR

The following is a compilation of sources of error that students may suggest:

1. The four forms of aluminum may not have had the same degree of purity.
2. The aluminum cylinder felt like it might have been hollow.
3. When we folded the aluminum foil, air might have been trapped inside.
4. Air might have been trapped by the aluminum shot when it was poured into the graduated cylinder.
5. The graduated cylinder may not have been sitting perfectly vertical in its plastic base, and therefore our volume readings may have been off.
6. When we added the various materials to the cylinder, water splashed onto the side of the cylinder, which changed the final volume.

The Graphical and Mathematical Relationship between Mass and Volume

THE EXPERIMENT

PURPOSE

To determine a graphical and mathematical relationship between the mass and volume of a given substance.

BACKGROUND

In this experiment, you will be given a sample of metal (copper, aluminum, zinc, etc.). Design an experiment to determine a graphical and mathematical relationship between the mass and volume of the substance.

PROCEDURE

NOTE: Consult safety information and obtain teacher approval before beginnning the experiment. Design an experiment to determine a graphical and mathematical relationship between the mass and volume of a given substance.

SAFETY INFORMATION

After reviewing your procedure, the instructor will discuss any safety precautions that are specific to your experiment.

QUESTIONS FOR FURTHER THOUGHT

1. Given that the density of water is 1.00 g/mL and the density of ethylene glycol (antifreeze) is 1.114 g/mL, if 100.0 mL of water is combined with 10.0 mL of ethylene glycol, what would be the density of the mixture?
2. Which element has the greatest density? Which element has the lowest density?
3. The density of toluene is 0.866 g/mL. What volume of toluene would have a mass of 15.00 g?

4 The Graphical and Mathematical Relationship between Mass and Volume

TEACHER'S NOTES

The intention of this lab is to introduce students to the concepts involved in analyzing data. In this experiment, students collect data on the mass and volume of various samples of a given material and are then asked to identify the mathematical and graphical relationships that exist between them.

Most students are familiar with the concept of density and can easily solve a density problem, but many students do not know how to analyze data to determine if variables are directly proportional, inversely proportional, have a linear relationship, and so on.

OPTIONS

1. This lab can be used in conjunction with the previous three labs on density if you are assigning each team one of the density-related activities.
2. The experiment also works well before beginning a unit on gas laws because it can be used to introduce the concepts of proportionality, proportionality constants, and graphical relationships.
3. If the entire class is performing this experiment, you can vary the metal being studied by each team.
4. If students have access to a graphing program such as Graphical Analysis by Vernier,[1] or have graphing calculators, they can obtain the equation of the best-fit line. Otherwise, students can try to estimate the equation of the best-fit line using graph paper.

NOTES

1. One of the metals used in the previous density labs (experiments 1–3) can also be used in this experiment. Therefore, if students have already calculated the density of the metal used in this lab during another experiment (or if this was done by another team), the value derived for the density during the other activity can be compared to the slope of the line obtained in this experiment.
2. Some students may only perform two trials, producing a graph with two points. This is a good time to point out the importance of multiple trials, especially when graphing data, as two points will always produce a linear relationship between the variables.
3. Often students are able to graph data easily but have trouble interpreting the meaning of the graph. You may want to spend time on this before students perform the activity or have the students try to discover the relationship on their own.

GRAPHICAL ANALYSIS OF THE DATA

You may want to suggest to the students that when they graph the data, they let the y-axis represent the mass of the sample, and the x-axis represent the volume. Then, when the students graph the data to obtain the equation of the best-fit line, the equation of the line will be:

Mass = (slope)(volume) + y-intercept

The y-intercept should be 0 (since a sample of metal with 0 mass will have 0 volume), and therefore the equation can be rewritten as:

$$\frac{\text{Mass}}{\text{Volume}} = \text{Slope}$$

and therefore, the slope of the line should also represent the density of the sample. (The calculated value for density and the slope of the best-fit line may vary significantly if the y-intercept of the best-fit line is not close to 0). If students reverse mass and volume on the graph, the slope of the line will approximate the inverse of the density.

MATHEMATICAL ANALYSIS OF THE DATA

Students who have studied proportionality and proportionality constants will realize that: (1) if two variables are directly proportional, their ratio will produce a constant, or (2) if two variables are inversely proportional, their product will produce a constant. These students may decide to analyze the data by determining if the ratio or product of each data set produces a constant. They should discover that the ratio of mass to volume produces a fairly consistent number, thereby showing that the two variables are directly proportional. Students should realize that this constant is known as density.

TIME

The amount of time needed for students to design and execute their own experiment can vary depending on the level and creativity of the class, their familiarity with designing their own investigations, and the complexity of the procedure they develop.

Average Time for This Experiment

Time needed to design the experiment: 15–20 minutes
Time needed to run the experiment: 30 minutes

TEAMS

Teams of two work well.

MATERIALS (PER TEAM)

Require the use of goggles, gloves, and aprons/lab coats. The following supplies should be readily available for the students:

- Sample of metal (aluminum shot, zinc pieces, copper shot, etc.)
- 50-mL or 100-mL graduated cylinder
- Weighing paper or weighing dishes
- Funnel and filter paper (if students will be draining the water from the metal shot)
- Balance

SAMPLE PROCEDURES

Students typically measure the mass of several samples and determine the volume of each sample through water displacement. Then they graph the data and determine the equation of the best-fit line. They should also try to manipulate the data to obtain a mathematical relationship between the variables.

SAFETY PRECAUTIONS

Discuss with the students any safety precautions that are specific to the procedure they developed. Material Safety Data Sheets for each substance used should be reviewed and made available to the students.

COMMON MISCONCEPTIONS, PROCEDURAL AND CALCULATION ERRORS

1. The most common problem for the students is how to analyze the data (unless they are familiar with proportional relationships and proportionality constants or using graphs to determine the relationship between the variables).
2. Students may reuse part or all of a sample in another trial, which will change the mass of the sample due to the wet BBs.
3. Students may use a beaker rather than a graduated cylinder to measure the volume of the BBs.

LAB REPORT

There are several alternatives to writing a formal lab report given in the introduction. If students are writing their own lab report it should include:

I. Title
II. Purpose
III. Materials
IV. Procedure
V. Data
VI. Calculations and/or graph
VII. Conclusion (including a discussion of sources of error)
VIII. Answers to Questions (optional)

ANSWERS TO THE QUESTIONS

1. 100.0 mL H_2O × 1.00 g/mL = 100.0 g H_2O
   10.0 mL ethylene glycol × 1.114 g/mL = 11.14 g ethylene glycol
   Total mass of solution = 111.14 g
   Total volume of solution = 110 mL
   Density of solution = 111.14 g / 110 mL = 1.01 g/mL
2. Osmium has the greatest density, 22.5 g/mL. Hydrogen has the lowest density, 8.4×10^{-5} g/mL at 1 atm and 20°C.
3. $\frac{15.00\text{ g}}{0.866\text{ g/mL}} = 17.3\text{ mL toluene}$

REFERENCE

1. Vernier Software, 8565 S.W. Beaverton-Hillsdale Hwy., Portland, OR 97225–2429 (503–297–5317).

FOR FURTHER READING

Wieczorek, L. *The Science Teacher.* May 1995, 61, pp. 44–45.

4 The Graphical and Mathematical Relationship between Mass and Volume

SAMPLE LAB REPORT

PURPOSE

To determine a mathematical relationship between the mass and volume for a given substance.

MATERIALS

- Safety equipment: goggles, apron, gloves
- Aluminum shot
- Weighing dish
- Balance
- 100-mL graduated cylinder

PROCEDURE

1. Determine the mass of a sample of aluminum shot.
2. Pour 50.0 mL of water into the graduated cylinder.
3. Add the sample of aluminum shot to the cylinder.
4. Determine the difference in volume in the cylinder. Record this volume as the volume of the shot.
5. Repeat steps 1–4 with new samples of aluminum shot.

DATA

SAMPLE	MASS (g)	VOLUME (mL)
1	5.81	2.1
2	14.16	5.3
3	38.33	14.4

CALCULATIONS AND GRAPH

1. For two variables to be inversely proportional, their product would produce a constant. For these values:

Sample 1: (5.81 g)(2.1 mL) = 12 g·mL

Sample 2: (14.16 g)(5.3 mL) = 75 g·mL
Sample 3: (38.33 g)(14.4 mL) = 552 g·mL

2. For two variables to be directly proportional their ratio would produce a constant. For these values:

Sample 1: $\frac{5.81\ \text{g}}{2.1\ \text{mL}} = 2.8\ \text{g/mL}$

Sample 2: $\frac{14.16\ \text{g}}{5.3\ \text{mL}} = 2.7\ \text{g/mL}$

Sample 3: $\frac{38.33\ \text{g}}{14.4\ \text{mL}} = 2.66\ \text{g/mL}$

3. See graph below. The equation of the best-fit line is shown.

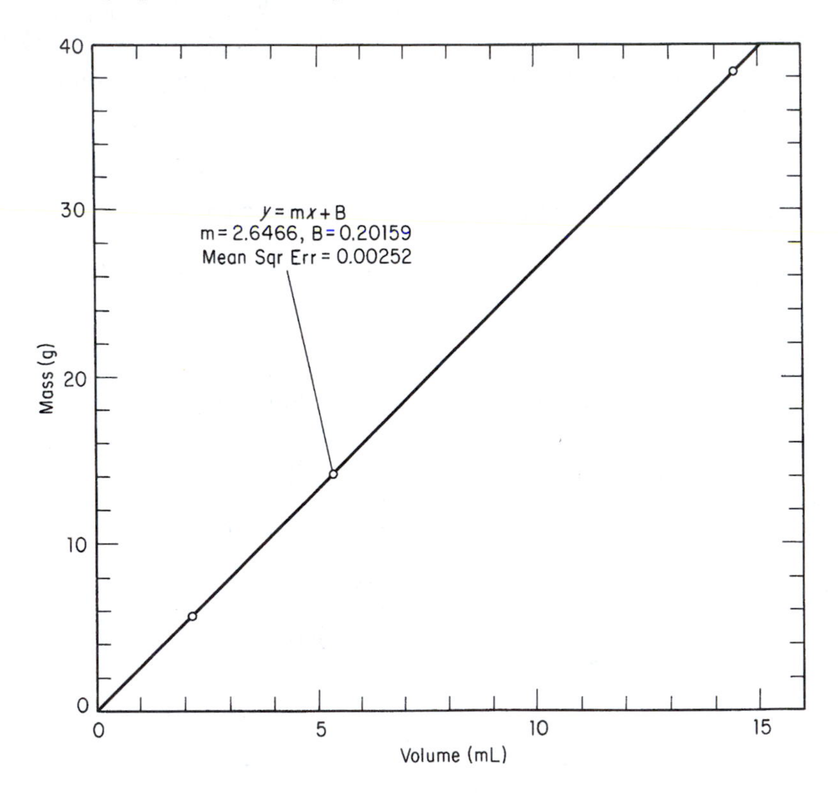

CONCLUSION

Mass and volume appear to be directly proportional for two reasons: (1) the ratio of the two variables produces a constant: (2.8 g/ml, 2.7 g/mL, 2.66 g/mL) and (2) the equation of the best-fit line is: mass = (2.65)(volume) + 0.2016

Since the y-intercept is so close to 0, the equation could be rewritten as mass = (2.65)(volume) or mass/volume = 2.65, again showing that the ratio of mass to volume produces a constant. The ratio shown in #1 above is approximately equal to the slope given on the graph. The values would be equal if the y-intercept on the graph was zero. Ideally, the y-intercept would be zero except for the various sources of error.

SOURCES OF ERROR

The following is a compilation of sources of error that students may suggest:

1. The metals are probably not pure because they have been used by other teams.
2. Water may have splashed out of the cylinder when we added the shot.
3. The metals may still be wet from another class.
4. Wind currents in the room may have affected the electronic balance.
5. We may have misread the graduated cylinder.

5 Determining the Density of Carbon Dioxide

THE EXPERIMENT

PURPOSE

To determine the density of carbon dioxide.

BACKGROUND

Carbon dioxide gas is released when an effervescent tablet such as Alka Seltzer™ is placed in water according to the equation:[1]

$$3NaHCO_{3(s)} + H_3C_6H_5O_{7(s)} + H_2O_{(l)} \rightarrow C_6H_5O_7Na_{3(aq)} + 4H_2O_{(l)} + 3CO_{2(g)}$$

Design an experiment to calculate the density of carbon dioxide. Check with the instructor to determine the number of tablets available for each team.

PROCEDURE

NOTE: Consult safety information and obtain teacher approval before beginning the experiment. Design an experiment to calculate the density of carbon dioxide gas.

SAFETY INFORMATION

1. Do not perform the reaction in a sealed container.
2. After reviewing your procedure, the instructor will discuss any safety precautions that are specific to your experiment.

QUESTIONS FOR FURTHER THOUGHT

1. What is the confirming test for carbon dioxide?
2. What is dry ice? What is the main purpose of dry ice?
3. What percentage of the atmosphere is carbon dioxide?
4. Why is carbon dioxide used in fire extinguishers?

REFERENCE

1. Faulkner, S. P. *The Science Teacher. 1993,* 60(1), 26–29.

5 Determining the Density of Carbon Dioxide

TEACHER'S NOTES

This experiment can produce a wide variety of procedures and methods and can be assigned while students are studying density or gases. If the lab is completed while the students are studying density, it is recommended that all students perform this activity in addition to one of the previous activities involving the density of metals.

To complete this lab it is helpful if students are familiar with the technique of capturing a gas by water displacement.

OPTIONS

Students often initially design an experiment that doesn't work as well as they anticipate. Review students' procedures to ensure that they are risk-free, and then allow students to find their own mistakes, to make this lab a good learning experience.

NOTES

1. Students may need to do one or two practice trials before beginning the experiment to test their setup and procedure.
2. Allow two to three tablets per team. You may want to tell the class that each team will be allotted three tablets and let the students decide how to divide them between trial runs and the actual experiment.
3. Effervescent tablets can be purchased through most chemical supply companies.
4. The volume of gas produced by the tablet can vary depending on the amount of water used in the reaction vessel, the amount of time given for the reaction to be completed, and whether the students agitate the flask to displace more carbon dioxide. Each tablet can produce between 150 mL and 250 mL of gas.
5. Allow two lab periods for the completion of this experiment.
6. You should insert the bent glass tubing into the rubber stoppers with glycerin rather than the students.
7. Allow the students approximately 40 minutes to decide on a procedure, gather their materials, and perhaps perform a trial run. Then provide a location where the students can keep their supplies until the next lab period.
8. Many electronic balances will not have the capacity to measure the mass of a flask or gas collecting bottle when it is full of water. You may need to have some triple beam balances available as well. A balance with a precision of ±0.001 g works best, although a balance with a precision of ±0.01 g can be used if the sample of carbon dioxide is large enough.
9. The density of carbon dioxide is approximately 0.002 g/mL at room temperature and room pressure.

TIME

The amount of time needed for students to design and execute their own experiment can vary depending on the level and creativity of the class, their familiarity with designing their own investigations, and the complexity of the procedure they develop.

Average Time for This Experiment

Time needed to design the experiment: 40 minutes
Time needed to run the experiment: 40 minutes

TEAMS

Teams of three work well.

MATERIALS (PER TEAM)

Require the use of goggles, gloves, and aprons/lab coats. The following supplies should be readily available for the students:

- Effervescent tablets (2–3 per team)
- Pneumatic trough
- Rubber tubing (to connect glass tubing and pneumatic trough)
- Pinch clamps
- Wide-mouth bottles
- Flasks (various sizes)
- Stopper to fit flask with bent glass tubing inserted
- Stopper to fit test tube with bent glass tubing inserted
- Gas measuring tube
- Stopper to fit gas measuring tube
- Test tubes
- 100-mL graduated cylinder
- Glass plate
- Balance with a precision of ±0.001 g. (A scale with a precision of ±0.01 g can be used, but students that collect only a small sample of gas will not be able to calculate the mass of the carbon dioxide.)

SAMPLE PROCEDURES

1. Most students will collect the gas by water displacement. They may react the tablet in a flask or test tube and capture the CO_2 in gas-collecting bottles, flasks, test tubes, gas measuring tubes, and so on. To determine the volume of the gas students may either read the volume directly (if using a graduated container such as a gas measuring tube) or compare the volume of the container when full to the volume of water remaining in the container. The predominant difference among the procedures is with the measurements taken to calculate the mass of the gas. Common methods include:

   A. Measuring the mass of the empty gas collecting container versus the mass of the container when it is full of carbon dioxide. The students will subtract, believing they have determined the mass of carbon dioxide, forgetting that the original container was not really empty but was filled with air.

   B. Measuring the mass of the gas collecting container when it is full of water versus the mass of the container when it contains carbon dioxide gas. Some students will then subtract the two values, believing they have calculated the mass of carbon dioxide gas. These students forget to account for the loss of mass due to the displacement of some water. The density value that they calculate will be close to 1.00 g/mL, the density of water. When students obtain this value, they should realize that it doesn't make sense. Students using this method and making this initial error in their calculations may believe that they have to repeat the entire experiment, when actually they can determine the mass of water lost through the following calculation: volume of carbon dioxide gas collected = volume of water lost = mass of water lost (see Sample Lab Report).

C. Comparing the mass of the dry tablet, reaction vessel, and water to the mass of the reaction vessel, water, and dissolved tablet. The difference in mass equals the mass of carbon dioxide produced from the reaction. Students using this method will need to collect all of the gas produced.

2. Some students may know that carbon dioxide is more dense than air and decide to direct the gas (via the rubber tubing) into a container and wait a few minutes for the gas to displace the air. These students will obtain the volume of the gas by measuring the volume of the bottle, but then have difficulty determining the mass of the gas. Students will probably compare the mass of the container when "empty" to the mass of the container when it is presumed to be full of CO_2.

SAFETY PRECAUTIONS

NOTE: Also refer to the safety information given with the experiment. Discuss with the students any safety precautions that are specific to the procedure they developed. Material Safety Data Sheets for each substance used should be reviewed and made available to the students.

COMMON MISCONCEPTIONS, PROCEDURAL AND CALCULATION ERRORS

NOTE: Several common errors are mentioned in Sample Procedures.

1. Some students may believe that they must capture all of the gas produced by the tablet, and they may completely disregard a particular trial if the container becomes filled with carbon dioxide while more gas is being produced. This may be a good time to remind students that an earlier lab showed that density is an intensive property. (NOTE: It *is* necessary for students to collect all of the gas produced by the tablet if they are determining the mass of the gas by comparing the mass of the tablet before the reaction [with the container and water] to the mass of the container and reacted tablet [see Sample Procedure 1C].)

2. Students might collect the gas in a flask and estimate the volume of gas produced using the graduations on the flask. This method is not as accurate as collecting the CO_2 in a gas-collecting tube or an inverted graduated cylinder.

3. Students often begin collecting the gas immediately rather than let-

ting it bubble into the trough for a few seconds to allow air to pass through the tubing.

4. Students may assume that all tablets have the same mass and will produce the same volume of gas.

LAB REPORT

There are several alternatives to writing a formal lab report given in the introduction. If students are writing their own lab report it should include:

I. Title
II. Purpose
III. Materials
IV. Procedure
V. Data
VI. Calculations
VII. Conclusion (including a discussion of sources of error)
VIII. Answers to Questions (optional)

ANSWERS TO THE QUESTIONS

1. When carbon dioxide is bubbled through a solution of calcium hydroxide (limewater), a precipitate of calcium carbonate is produced.
2. Dry ice is solid carbon dioxide. It is used as a refrigerant.
3. Carbon dioxide is approximately 0.036 mole percent of the atmosphere.
4. Carbon dioxide is used in fire extinguishers because it is more dense than air and does not support combustion. Because it is more dense than air, it settles around the fire, displacing the oxygen and extinguishing the flames.

5 Determining the Density of Carbon Dioxide

SAMPLE LAB REPORT

PURPOSE

To determine the density of carbon dioxide.

MATERIALS

- Safety equipment: goggles, apron, gloves
- Effervescent tablet
- Pneumatic trough
- 250-mL flask with rubber stopper and bent glass tubing
- Rubber tubing
- Gas-collecting (wide mouth) bottle
- 100-mL graduated cylinder
- Glass plate
- Pinch clamp
- Balance

PROCEDURE

1. Attach one end of the rubber tubing to the pneumatic trough.
2. In the middle of the tubing, shut the tubing using the pinch clamp.
3. Fill the trough about 3/4 full of water.
4. Attach the other end of the tubing to the bent glass in the stopper.
5. Fill the wide-mouth bottle with water. Pour the water into a graduated cylinder to determine the volume of the wide-mouth bottle. Record this volume.
6. Refill the wide mouth bottle, cover the bottle with a glass plate and determine the mass of the bottle, water, and glass plate.
7. Invert the wide-mouth bottle into the trough. Place the bottle over the opening in the bottom of the trough. Remove the glass plate.
8. Pour approximately 125 mL of water into the flask. (The following step must be done quickly and carefully, so pressure does not increase substantially within the flask.)
9. Put the effervescent tablet in the flask, insert the stopper into the flask, and immediately remove the pinch clamp from the rubber tubing.
10. When the reaction is complete, use the glass plate to invert the wide-mouth bottle, being careful not to lose any remaining water.
11. Measure the mass of the wide mouth bottle, remaining water (if any), glass plate, and CO_2.
12. Determine the volume of the water (if any) remaining in the wide-mouth bottle.

DATA

Volume of gas-collecting bottle: 295.5 mL

Mass of gas-collecting bottle when full of water (with glass plate): 537.879 g

Mass of gas collecting bottle, remaining water, and CO_2: 284.138 g

Volume of water remaining in gas-collecting bottle: 34.9 mL

CALCULATIONS

1. Volume of gas produced = volume of gas collecting bottle − volume of remaining water
   = 295.5 mL − 34.9 mL
   = 260.6 mL
2. Volume of water displaced = volume of gas produced = mass of water displaced
   = 260.6 mL CO_2
   = 260.6 g H_2O
3. Mass of bottle at the end with displaced water added = 284.138 g + 260.6 g
   = 544.7 g

NOTE: The most common mistake made by students in their calculations is to forget to account for the loss of mass due to the loss of water in the bottle. Students need to include the mass lost due to lost water in their calculation.

4. Mass of gas = 544.7 g − 537.879 g
   = 6.8 g

$$\text{Density of carbon dioxide} = \frac{\text{Mass}}{\text{Volume}} = \frac{6.8\text{ g}}{260.6\text{ mL}} = 0.026\text{ g/mL}$$

CONCLUSION

The density of carbon dioxide is 0.026 g/mL. (NOTE: The data did not produce an answer that is close to the accepted value but is representative of what students might obtain.)

SOURCES OF ERROR

The following is a compilation of sources of error that students may suggest:

1. Some water or carbon dioxide may have been accidentally released when removing the bottle from the trough.
2. The outside of the bottle was wet when we measured its mass with the CO_2.
3. We didn't collect all of the gas produced by the tablet. (This would not be a source of error with the procedure included in the Sample Lab Report.)
4. Some of the carbon dioxide gas may have dissolved in the water.
5. There may have been some air collected in the bottle. We can't be sure it was 100% CO_2.

6 Determining the Percentage Composition of a Mixture

THE EXPERIMENT

PURPOSE

To determine the percentage composition of a mixture.

BACKGROUND

You will be given a mixture of two substances. (Your instructor will tell you the components of the mixture). The purpose of the experiment is to determine what percentage of the mixture's mass is due to each of the various substances.

PROCEDURE

NOTE: Consult safety information and obtain teacher approval before beginning the experiment. Design an experiment to determine the percentage composition of the mixture.

SAFETY INFORMATION

After reviewing your procedure, the instructor will discuss any safety precautions that are specific to your experiment.

QUESTIONS FOR FURTHER THOUGHT

1. If all of the students in the class received a sample of the same mixture, should each person (or team) obtain the same result for its percentage composition? Why?
2. If the purpose of the experiment was to determine the percentage composition of a compound, rather than a mixture, should each person (or team) obtain the same result for the percentage composition? Why?
3. If the purpose of the experiment was to determine the percentage composition of a solution, rather than a mixture, should each person (or team) obtain the same result for the percentage composition? Why?
4. Does your experiment use physical changes or chemical changes to separate the components of the mixture?

REFERENCE

Weast, R.C., ed., *Handbook of Chemistry and Physics*, CRC Press, Cleveland, OH, 1976.

6 Determining the Percentage Composition of a Mixture

TEACHER'S NOTES

In this lab, students are given a mixture of two materials (a solid that is insoluble in water and a solid that is soluble in water) and are asked to determine the percentage composition of the mixture (i.e., what fraction of the mixture's mass is due to each of the components).

This experiment can be used to (1) introduce the concept of percentage composition when studying the stoichiometry of chemical formulas, or (2) reinforce the characteristics of mixtures and some physical changes that can be used to separate the components of a mixture.

OPTIONS

1. Different teams can be given samples of the same mixture to show that the composition of the mixture will vary since it is heterogeneous.
2. Different teams can be given different mixtures and compare their techniques.
3. Students can write their final percentage composition values on the blackboard to compare the values among teams. This can be useful if the purpose of the experiment is to discuss the properties of mixtures and their heterogeneity.

NOTES

1. Some materials that can be used include

   Salt and sand
   Sugar and sand
   Sodium bicarbonate (baking soda) and sand
   Salt and marble chips

2. Clean white sand can be purchased from most chemical supply companies.
3. Determine the mass of the components of the mixture before combining them, so the correct percent composition is known.
4. Provide each team with a small scoop of the dry mixture (approximately 10–15 g or 1 tablespoon).
5. If students are not familiar with the technique of washing a precipitate, this is a good experiment in which to introduce the concept.
6. Discourage students from using large quantities of water as it can significantly increase the amount of time needed to complete the experiment.

TIME

The amount of time needed for students to design their own experiment can vary depending on the level and creativity of the class, their familiarity with designing their own investigation, and the complexity of the procedure they develop.

Average Time for This Experiment

Time needed to design the experiment: 20 minutes

Time needed to run the experiment: 30–60 minutes and some time another day

TEAMS

Teams of two to three work well.

MATERIALS (PER TEAM)

Require the use of goggles, gloves, and aprons/lab coats. The following supplies should be readily available for the students:

- Mixture: 5–10 g per team
- 250-mL or 400-mL beakers
- Heat source
- Watch glasses to cover the beakers (to avoid spattering when boiling the solution)
- Rubber policeman (to remove sand from the beaker)
- Stirring rod
- Filter paper
- Funnel
- Wire gauze
- Beaker tongs or insulated gloves
- Distilled water

SAMPLE PROCEDURES

Before the students begin designing their procedure, make it clear that they will have time another day to finish the lab.

Students will probably design a procedure in which they initially dissolve the soluble component (sugar, salt or baking soda) and filter the insoluble component (sand or marble chips). However, the method used by the students to calculate the percentage composition can vary from team to team. Some teams will measure the mass of both components independently, while other teams will only measure the mass of the mixture and one of the components and then subtract to obtain the mass of the other component.

Methods commonly used include:

1. Boil the water solution to dryness and determine the mass of the soluble product. Allow the filter paper to dry overnight and calculate the mass of the insoluble product.
2. Boil the water solution to dryness and determine the mass of the soluble product. Subtract this mass from the original mass of the mixture to obtain the mass of the insoluble product.
3. Allow the filter paper to dry overnight and measure the mass of the insoluble product. Subtract this value from the mass of the mixture to obtain the mass of the soluble product.

Other variations that students may use include:

1. Boiling the mixture of water, soluble, and insoluble components (before filtering) to help ensure that all of the soluble product dissolves.
2. Using the rubber policeman to scrape the wet sand from the filter paper into an evaporating dish, followed by heating the sand to evaporate any water (rather than allowing it to dry overnight).
3. Pouring the water and soluble product from one beaker into another

(leaving the sand behind in the original beaker), followed by several rinses of the insoluble product with distilled water. This procedure eliminates the need for filtering.

SAFETY PRECAUTIONS

Discuss with the students any safety precautions that are specific to the procedure they develop. Material Safety Data Sheets for each substance used should be reviewed and made available to the students.

For teams that decide to boil the water solution to dryness, some suggestions are to (1) cover the beaker with a large watch glass to avoid spattering the soluble component, (2) reduce the heat if spattering does occur, and (3) remove the beaker from the hot plate (or turn off the Bunsen burner) as soon as it appears that most of the water has been evaporated.
NOTE: Boiling the solution with the soluble product will work safely when small amounts (10–15 g) of the mixture have been provided. If students boil solutions containing significantly larger quantities of salt, safety issues can become a factor. Additionally, the smaller the relative amount of salt in the mixture, the lower the safety risks during boiling.

COMMON MISCONCEPTIONS, PROCEDURAL AND CALCULATION ERRORS

1. Students may think that they need to include the mass of the water in their calculations.

2. Some students may think they can separate the components by heating the original dry mixture until the salt or sugar melts. Remind them that salt and sugar both have very high melting points. This procedure should not be permitted.

3. Students may not be sure about the quantity of water required to dissolve all of the salt or sugar. Remind them that salt and sugar are very soluble, or refer them to a solubility curve in the text if it shows the solubility of the substance in question or refer them to the *Handbook of Chemistry and Physics.*

LAB REPORT

There are several alternatives to writing a formal lab report given in the introduction. If students are writing a formal lab report it should include:

I. Title
II. Purpose
III. Materials
IV. Procedure
V. Data
VI. Calculations
VII. Conclusion (including a discussion of sources of error)
VIII. Answers to Questions (optional)

ANSWERS TO THE QUESTIONS

1. No, each team would not be expected to get the same results for the percentage composition of the mixture. There are two main reasons for this:

(a) inaccuracies in the measurements and other sources of error will cause the results to fluctuate from team to team, and (b) the teams were given a heterogeneous mixture to analyze and therefore different samples from the same mixture would not have precisely the same composition.

2. If the teams were analyzing a compound rather than a mixture, the results should be identical except for any errors or inaccuracies in the measurements, mistakes in the procedure, and so on. This is because compounds are homogeneous.
3. Solutions are homogeneous mixtures and therefore each team should obtain identical results except for any errors or inaccuracies in the measurements, mistakes in the procedure, and so on.
4. Mixtures can be separated using physical changes. The physical changes used in this experiment include dissolving, filtering, and boiling.

6 Determining the Percentage Composition of a Mixture

SAMPLE LAB REPORT

PURPOSE

To determine the percentage composition of a mixture composed of salt and sand.

MATERIALS

- Safety equipment: goggles, apron, gloves
- Mixture
- 400-mL beaker
- Hot plate
- 250-mL beaker
- Watch glass (to fit 250-mL beaker)
- Stirring rod
- Beaker tongs
- Rubber policeman
- Distilled water
- Filter paper
- Funnel

PROCEDURE

1. Measure the mass of the mixture of sand and salt.
2. Place the mixture in a 400-mL beaker.
3. Add approximately 100 mL of water to the mixture and stir for 3 minutes.
4. Measure the mass of the filter paper.
5. Fold the filter paper and place it in the funnel.
6. Measure the mass of the 250-mL beaker and watch glass.

7. Using the rubber policeman, pour the solution and sand into the filter paper, allowing the filtrate to flow into the 250-mL beaker.
8. After the solution and sand has passed into the filter paper, pour 20 mL of distilled water into the filter paper to wash the sand and rinse any remaining salt water into the beaker.
9. Cover the beaker with the watch glass (making sure that steam will be able to escape) and heat gently until all of the water has boiled away. Immediately remove the beaker from the hot plate and allow it to cool.
10. Place the beaker of salt in an area where it will not be disturbed.
11. Remove the filter paper from the funnel and place it in an area where it will not be disturbed.
12. The next day, measure the mass of the filter paper and sand as well as the mass of the beaker and salt.

DATA

Mass of mixture: 10.32 g
Mass of filter paper: 1.02 g
Mass of 250-mL beaker and watch glass: 100.97 g
Mass of filter paper and sand after drying: 7.79 g
Mass of 250-mL beaker, watch glass, and salt after boiling: 104.38 g

CALCULATIONS

Mass of sand: 7.79 g − 1.02 g = 6.77 g
Mass of salt: 104.38 g − 100.97 g = 3.41 g
% Sand in mixture: 6.77 g sand/10.32 g mixture = 0.656 × 100 = 65.6% sand
% Salt in mixture: 3.41 g salt/10.32 g mixture = 0.330 × 100 = 33.0% salt

NOTE: The percentages in this experiment will not add up to 100% because the students measured the mass of salt and sand independently. Teams that determine the mass of the salt or sand by subtracting the mass of the other from the mass of the mixture will obtain percentage values that add up to 100%.

CONCLUSION

The percentage composition of the mixture is 65.6% sand and 33.0% salt.

SOURCES OF ERROR

The following is a compilation of sources of error that students may suggest:

1. Because the quantity of salt in the mixture is unknown, it was difficult to determine if all of the salt had dissolved in the water. The solution may have been saturated and some excess salt remained with the sand.
2. Despite washing the sand with distilled water, some salt may have remained behind on the sand or filter paper.

3. Some salt spattered out of the beaker during boiling.
4. The filter paper may not have been completely dry the next day.
5. Sodium chloride absorbs moisture from the air and may have absorbed water during the night.
6. Contaminants in the tap water may have remained behind with the salt after boiling.

7 Heat of Fusion of Ice

THE EXPERIMENT

PURPOSE

To determine the heat of fusion of ice.

BACKGROUND

The heat of fusion for a given substance is the quantity of heat needed to melt a given amount of that substance at a constant temperature. The Standard International (SI) unit for heat of fusion is kiloJoules per mole (kJ/mol). Heat of fusion can also be measured in kJ/g, kcal/mol, cal/g, or kcal/g. Check with your instructor regarding which unit to use.

In this experiment, you will calculate the heat of fusion of ice through the use of calorimetry. In calorimetry,

1. The substance that is undergoing a change in heat content is placed in a calorimeter containing a known quantity of water. (For this experiment, you will be choosing the container that will be used as the calorimeter.)
2. It is assumed that the quantity of heat lost or gained by the substance is equal to the quantity of heat lost or gained by the water.
3. The amount of heat lost or gained by the water can be determined using the formula:

   $\Delta H = \Delta t \times m \times c$

   [where]

   ΔH = the change in heat quantity of the water in the cup,
   Δt = the change in temperature of the water in the cup
   m = mass of the water undergoing the temperature change, and
   c = specific heat of water (The specific heat for water = 1.00 cal/g·°C. = 4.18 J/g·°C)

In this experiment, you will decide the quantity of water to have in the calorimeter, the initial and final temperature of the water, the quantity of ice to use, and what type of container to use for a calorimeter.

MATERIALS

- Safety equipment: goggles, apron, gloves
- 400-mL beaker
- 100-mL graduated cylinder
- Thermometer
- Calorimeter
- Tongs
- Hot plate (or Bunsen burner with ring stand, iron ring, and wire gauze)
- Ice cubes

PROCEDURE

NOTE: Consult safety information before beginning the experiment.

Before beginning the experiment, decide how much water to use in the calorimeter, what the initial and final temperature of the water should be, the amount of ice to be used, and what container should be used as the calorimeter.

1. If you will be using tap water in the calorimeter, the desired volume can be added directly to the chosen container. If you decide to heat or cool the water before beginning, the temperature should be adjusted in a beaker and then the desired volume poured in the calorimeter.
2. Record the temperature of the water in the calorimeter.
3. Immediately after recording the temperature of the water, place the desired number of ice cubes in the calorimeter.
4. Monitor the temperature of the water as you gently stir the mixture until it has reached the chosen final temperature.
5. After recording the final temperature of the mixture, use the tongs to remove any unmelted ice.
6. Determine the final volume of the water in the calorimeter.

SAFETY INFORMATION

1. Stir the mixture carefully with the thermometer. If the thermometer breaks, alert the instructor immediately.
2. Follow all safety precautions when using a Bunsen burner.
3. After reviewing your procedure, the instructor will discuss any safety precautions that are specific to your experiment.

QUESTIONS FOR FURTHER THOUGHT

1. Is fusion (melting) an endothermic or exothermic process?
2. Is a high heat of fusion indicative of strong attractive forces be-

tween the particles in the substance or weak attractive forces between the particles in the substance?

3. What are some ways this experiment could be improved to minimize any potential sources of error?

7 Heat of Fusion of Ice

TEACHER'S NOTES

This lab is intended to be an introduction to the concepts of calorimetry, and therefore a general procedure is included in the experiment, although students do make several decisions regarding certain aspects of the lab. There are several other experiments that involve the use of calorimetry (experiment 8, Heat of crystallization of wax; experiment 10, Comparing the specific heat of metals and nonmetals; experiment 16, Calculating the heat of solution; and experiment 20, Calculating the quantity of heat released from The Heat Solution™ handwarmer) where the students are not given a procedure to follow.

If, however, you want to have the students develop their own procedures, you could pose the problem to the students and perhaps introduce some of the concepts involved in calorimetry. The procedure is provided as an option.

The students are given some flexibility regarding the units that can be used for heat of fusion. Although the SI unit for heat of fusion is kJ/mol, some textbooks cover the concepts of heat and kinetic energy before they introduce the concept of moles. The decision regarding which unit to use is left to your discretion.

OPTIONS

1. You can have the students follow this procedure to introduce the concepts involved in calorimetry or allow the students to design their own experiment. If students are designing their own procedure, they should be familiar with the equation used to calculate the change in heat.

2. Even if the students are following the written procedure, they can incorporate their own variations, perhaps performing several trials and (a) vary the initial volume of water in the cup, (b) vary the container used for the calorimeter, or (c) use distilled water instead of tap water. Students could then compare the results and decide which trial was the most accurate.

3. After completing this experiment, the students could be given a sample of ice and asked to calculate its mass by designing and executing their own calorimetry experiment.[1]

NOTES

1. Accepted values for the heat of fusion of ice are 80 cal/g or 6.01 kJ/mol.

2. When deciding what volume of water and calorimeter to use in step 3 of the procedure, the students should consider the following:

A. Allowing room in the calorimeter for the addition of the ice,
B. A large volume of water means a smaller ΔT (given the same amount of ice) and perhaps less accuracy,

C. Using a very small volume of water may mean that some of the ice won't be submerged and will absorb heat from the air rather than the water. Remember the assumption is that all of the heat absorbed by the ice is lost from the water, and

D. An insulated container such as a polystyrene cup would reduce the exchange of heat between the water and the air.

3. When deciding the initial and final temperature of the water, students may decide that a large ΔT would be better than a smaller value and may heat the water to near boiling. If they also decide to drop the temperature to 2° or 3°C, a large amount of ice will be required.

4. If only a limited supply of ice is available, you can specify the number of ice cubes available for each team.

5. A volume of water between 75.0 mL and 150.0 mL in the calorimeter works fine, depending on the size of the cup or beaker used.

6. Students may experience trouble with the calculations if this is their first experiment involving the concepts of calorimetry. You may want to review the following concepts before beginning the experiment:

A. The ice will absorb heat from the water while it melts, causing the temperature of the water in the cup to decrease.

B. The quantity of heat absorbed by the ice is *assumed* to be equal to the quantity of heat lost by the water in the cup.

C. To calculate the quantity of heat lost by the water in the cup, the ΔH equation given in the background portion of the lab is used. Specify to the students whether you want the calculations performed using calories or joules and grams or moles. Emphasize that the ΔH equation is in terms of the liquid water in the cup.

7. The amount of information that you want to review with the students or provide depends on the level of the students and the degree of independence that you want them to have on this experiment.

TIME

The amount of time needed for students to design and execute their own experiment can vary depending on the level and creativity of the class, their familiarity with designing their own investigations, and the complexity of the procedure they develop.

Average Time for This Experiment

Pre-lab: 10–15 minutes (more time may be needed if you want to review all of the concepts and equations involved).

Lab: 20–30 minutes (students often have time for two trials).

NOTE: Allow more time if the students are designing their own experiment.

TEAMS

Teams of two work well.

MATERIALS (PER TEAM)

Require the use of goggles, gloves, and aprons/lab coats. The following supplies should be readily available:

- 400-mL beaker
- 100-mL graduated cylinder
- Thermometer
- Tongs
- Hot plate (or Bunsen burner with ring stand, iron ring, and wire gauze)
- Ice cubes (for a class of 20 students, you will need about 2–3 trays of ice cubes)
- Containers that could serve as a calorimeter, including Styrofoam cups, beakers, metal cups, and so on.

SAFETY PRECAUTIONS

Refer to the safety information given with the experiment. Discuss with the students any safety precautions that are specific to the procedure they develop. Material Safety Data Sheets for each substance used should be reviewed and made available to the students.

COMMON MISCONCEPTIONS, PROCEDURAL AND CALCULATION ERRORS

1. Students may forget that the density of water is 1.00 g/mL. This value is needed to perform the calculations.
2. Students often confuse temperature and heat. If students design their own experiment, they may decide that the change in temperature of the water in the calorimeter is equal to the heat lost by the water.
3. Students may mistakenly use the mass of the ice that melted for m in the ΔH equation. They should understand that the values used in the ΔH equation all refer to the initial amount of *liquid water* in the cup and calculate the change in heat of the liquid water. Then, it is assumed that the heat lost by the liquid water is equal to the heat gained by the ice.

LAB REPORT

There are several alternatives to writing a formal lab report given in the introduction.

If students are writing a formal lab report, it should include:

I. Title
II. Purpose
III. Data
IV. Calculations (calculating the heat of fusion of ice and percent error)
V. Conclusion (including a discussion of sources of error)
VI. Answers to Questions (optional)

Students will add materials and procedure if they are designing their own experiment.

ANSWERS TO THE QUESTIONS

1. Fusion is an endothermic process.
2. A high heat of fusion indicates that a large amount of energy is needed

to break the bonds or forces that are holding the substance in the solid phase. Therefore, a substance with a high heat of fusion has relatively strong attractive forces between its particles.

3. Improvements that could be made in the experiment might include (a) using distilled water in the cup and for ice cubes, (b) using a second Styrofoam cup as an extra insulator, and (c) checking the calibration of the thermometers.

REFERENCE

1. Plumsky, R. *Journal of Chemical Education.* 1996, 73, 451–454.

FOR FURTHER READING

Nheh, L.N., Orbell, J.D., Bigger, S.W. *Journal of Chemical Education.* 1994, 71, 793–795.

7 Heat of Fusion of Ice

SAMPLE LAB REPORT

PURPOSE

To determine the heat of fusion of ice.

DATA

Volume of water in Styrofoam cup: 100.0 mL
Mass of water in Styrofoam cup: 100.0 g
Initial temperature of water: 48.0°C
Final temperature of water: 4.0°C
Final volume of water in cup: 150.0 mL

CALCULATIONS

NOTE: Typically students do not consider the quantity of heat used to raise the temperature of the melted ice from 0°C to the final temperature of the water, so those calculations are not shown here. Occasionally, students will subtract this quantity of heat from the heat lost from the original sample of water.

Heat lost by water in cup (ΔH) $= \Delta T \times m \times c$
$= 44.0°\text{C} \times 100.0 \text{ g} \times 1.00 \text{ cal/g·°C}$
$= 4.40 \times 10^3 \text{ cal}$

Heat absorbed by the ice = heat lost by the water
$= 4.40 \times 10^3 \text{ cal}$

Mass of ice that melted = change in volume in the cup
= 50.0 g

Heat of fusion of ice = quantity of heat absorbed by the ice/mass of ice that melted
$= 4.40 \times 10^3$ cal/50.0 g
$= 88.0$ cal/g

Percent error = accepted value − experimental value/accepted value $\times 100$
$= 80 - 88.0/80 \times 100$
$= -10\%$

CONCLUSION

Based on the data, the heat of fusion of ice is 88.0 cal/g.

SOURCES OF ERROR

The following is a compilation of sources of error that students may suggest:

1. Some of the heat that was lost by the water may have been lost to the environment rather than absorbed by the ice.
2. Some water may have been left behind in the graduated cylinder when it was first poured into the cup.
3. Some of the water in the cup may have been on the ice when the ice was removed.
4. Some water may have been left on the tongs when the ice was removed.
5. Because ice floats, it may have absorbed some heat from the air rather than from the water in the cup.
6. Some of the warm water may have evaporated, changing the mass of the water in the cup.
7. The ice may not be pure, and impurities may absorb heat.
8. Dropping ice into the water can cause some water to splash out of the cup.
9. Some water was left behind in the cup when the water was transferred to the graduated cylinder for the final volume measurement.
10. The ice may contain air bubbles that may absorb heat from the water but will not affect the final volume measurement.
11. The thermometer may have been misread.
12. The thermometer may not be calibrated correctly.
13. Once the water in the cup was below room temperature, some water vapor from the room may have condensed into the cup.

8 Heat of Crystallization of Wax

THE EXPERIMENT

PURPOSE

To determine the heat of crystallization of wax.

BACKGROUND

The heat of crystallization is the quantity of heat released as a unit mass of a substance solidifies at constant temperature. The SI unit for heat of crystallization is kJ/mol, although kJ/g or cal/g can also be used. For this lab, the unit of kJ/g, J/g, or cal/g should be used.

In this experiment, you will be determining the heat of crystallization of wax through the process of calorimetry.

PROCEDURE

NOTE: Consult safety information and obtain teacher approval before beginning the experiment. The instructor will provide a sample of wax and indicate the mass of wax contained in the test tube. Design an experiment to determine the heat of crystallization of wax.

SAFETY INFORMATION

1. Initially melt the wax in a hot water bath. Do not directly heat the test tube of wax.
2. Do not place the thermometer in the wax.
3. Return the sample of wax to the instructor when the lab is completed.
4. After reviewing your procedure, the instructor will discuss any safety precautions that are specific to your experiment.

QUESTIONS FOR FURTHER THOUGHT

1. How does the heat of fusion for a given substance compare to the value for the heat of crystallization?
2. Is crystallization an endothermic or exothermic process?

8 Heat of Crystallization of Wax

TEACHER'S NOTES

In this experiment, students are asked to determine the heat of crystallization of wax using calorimetry.

The lab should be performed after students complete experiment 7, Heat of fusion of ice, as they must be familiar with the processes and calculations involved in calorimetry to complete this experiment.

OPTIONS

1. Provide students with the equations used in calorimetry.
2. Use various types of waxes and compare the values derived for heat of crystallization.
3. Place the test tubes in a hot water bath before the students arrive to decrease the time required to complete the lab.
4. Specify which unit to use for the heat of crystallization.

NOTES

1. The heat of crystallization of candle wax is about 60 cal/g.
2. The test tubes should contain 8.00–12.00 g of wax, depending on the size of the test tube (larger test tubes work well). The test tubes should be labeled with the mass of wax contained in the tube.
3. You can purchase candles from a supply company and cut the wax into pieces small enough to fit into the test tube.
4. The test tubes can be stoppered and reused for many years.
5. Emphasize to the students that the heat of crystallization is the quantity of heat released during the freezing process. This should provide a hint to the students that they should wait until crystallization begins before placing the test tube of melted wax into the calorimeter.
6. Students should decide how much water to place in the calorimeter before beginning the experiment. They should make sure that the wax is completed submerged, but should also realize that using a large volume of water will result in a small ΔT, which may be difficult to measure.
7. If students wait for crystallization to begin before placing the test tube in the calorimeter, they will eventually notice a small ball of solid wax forming in the bottom of the test tube. Students will need to hold the test tube in the air and carefully watch this area.

TIME

The amount of time needed for students to design and execute their own experiment can vary depending on the level and creativity of the class, their familiarity with designing their own investigations, and the complexity of the procedure they develop.

Average Time for This Experiment

Time needed to design the experiment: 20–30 minutes
Time needed to run the experiment: 30–40 minutes

TEAMS

Teams of two work well.

MATERIALS (PER TEAM)

Require the use of goggles, gloves, and aprons/lab coats. The following supplies should be readily available for the students:

Sample of wax in a test tube (labeled with the mass of the wax sample)
Heat source
400-mL beaker
Test tube holder
Thermometer
100-mL graduated cylinder
Styrofoam cup (8 or 16 ounce)
Other containers that could serve as a calorimeter (e.g., beakers, metal cups)
Students may also ask for various materials to cover the calorimeter and provide extra insulation. Materials could include wax paper, aluminum foil, and so on.

SAMPLE PROCEDURES

Students may use the following general procedure:

1. Melt the wax in a hot water bath. (Students should begin heating the water as soon as possible.)

2. Pour a given amount of water into the calorimeter. (Students should make sure that the wax will be beneath the surface of the water. Using the minimum amount of water needed to submerge the wax will provide the most accurate results as the value for ΔT will be larger and more measurable).

3. Remove the test tube of melted wax from the hot water bath and place it in the calorimeter. (Most students will immediately transfer the test tube of melted wax to the calorimeter, even though they should ideally wait for crystallization to begin. Immediate placement in the calorimeter results in additional heat being transferred to the water as the wax cools to its freezing point. This will produce a larger ΔT and greater heat of crystallization values than would be expected.)

SAFETY PRECAUTIONS

Refer to the safety information given with the experiment. Discuss with the students any safety precautions that are specific to the procedure they develop. Material Safety Data Sheets for each substance used should be reviewed and made available to the students.

Hot wax can cause painful burns; students should take appropriate precautions.

COMMON MISCONCEPTIONS, PROCEDURAL AND CALCULATION ERRORS

1. Students may immediately place the test tube of melted wax into the calorimeter rather than waiting for crystallization to begin.

2. Students will use the mass of the wax, rather than the mass of the wa-

ter, when initially calculating the quantity of heat gained by the water in the cup.

3. Students may simply use the calculated value for ΔH as the heat of crystallization and not convert to kJ/g, J/g, or cal/g.

LAB REPORT

There are several alternatives to writing a formal lab report given in the introduction. If students are writing their own lab report it should include:

I. Title
II. Purpose
III. Materials
IV. Procedure
V. Data
VI. Calculations
VII. Conclusion (including a discussion of sources of error)
VIII. Answers to Questions (optional)

ANSWERS TO THE QUESTIONS

1. For a given substance, the heat of fusion is numerically equal to the heat of crystallization. However, during fusion the heat is absorbed and during crystallization the heat is released.
2. Crystallization is exothermic. Evidence from the lab that would support this is that as the wax freezes (or crystallizes), the temperature of the water in the calorimeter increases.

8 Heat of Crystallization of Wax

SAMPLE LAB REPORT

PURPOSE

To determine the heat of crystallization of wax.

MATERIALS

Safety equipment: goggles, apron, gloves
Sample of wax (9.00 g)
Hot plate
Styrofoam cup (calorimeter)
400-mL beaker
Test tube holder
Thermometer
100-mL graduated cylinder

PROCEDURE

1. Fill the 400-mL beaker about 3/4 full with tap water.
2. Begin heating the water on the hot plate.

3. Place the test tube of wax in the calorimeter and add water to the cup until the wax is just submerged. Remove the test tube and pour the water into the graduated cylinder to determine the minimum amount of water needed. Record the volume of water and pour it back into the calorimeter.
4. Place the test tube of wax in the hot water bath and continue heating until the wax is completely melted.
5. Record the temperature of the water in the calorimeter.
6. Using the test tube holder, hold the test tube in the air until a small ball of solidified wax appears in the bottom of the test tube.
7. Immediately place the test tube in the calorimeter. Monitor the temperature of the water, recording the maximum temperature.
8. Return the sample of wax to the instructor.

DATA

Volume of water in calorimeter = 80.0 mL (80.0 g)
Mass of wax sample = 9.00 g
Initial temperature of the water = 22.0°C
Final temperature of the water = 28.0°C

CALCULATIONS

Heat gained by the water (ΔH) = 80.0 g × 6.0°C × 1.00 cal/g·°C = $4\bar{8}0$ cal
Heat lost by the wax sample = $4\bar{8}0$ cal
Heat of crystallization of wax = $4\bar{8}0$ cal/9.00 g = 53 cal/g

CONCLUSION

The heat of crystallization of wax is 53 cal/g.

SOURCES OF ERROR

The following is a compilation of sources of error that students may suggest:

1. The wax sample may not be exactly 9.00 g because it has been used by other teams.
2. The wax sample may be contaminated.
3. Water may have splashed into the test tube.
4. The test tube had hot water on the outside that may have released some heat into the calorimeter.
5. Some wax had crystallized before it was placed in the calorimeter, therefore some of the heat released during crystallization was lost to the air.
6. Some heat may have been transferred through the glass test tube to the air.
7. The specific heat of tap water may be not be the same value as it is for pure water (i.e., 1.00 cal/g·°C).

9 Determining the Specific Heat of a Metal

THE EXPERIMENT

PURPOSE

To determine the specific heat of a metal.

BACKGROUND

The specific heat (or specific heat capacity) of a metal is the quantity of heat energy needed to raise a given mass of a metal 1°C. This is equivalent to the quantity of heat energy released when the same mass of metal drops 1°C. The SI unit for specific heat is J/kg·K (although kcal/g·°C or cal/g·°C are also acceptable).

In this experiment you will be given a sample of metal and will use the process of calorimetry to determine its specific heat.

The procedure involves placing the metal in a beaker of boiling water to bring the metal to an initial temperature equal to the temperature of the boiling water. The hot metal will then be placed in a calorimeter containing a known volume of water at room temperature. (Your team will decide what volume of water to use in the calorimeter and what container to use as the calorimeter). The heat from the metal will be released into the cooler water until the metal and the water reach the same temperature.

MATERIALS

- Safety equipment: goggles, apron and gloves
- Sample of a known metal
- Calorimeter
- 400-mL beaker
- 100-mL graduated cylinder
- Thermometer
- Heat source
- Tongs

PROCEDURE

NOTE: Consult safety information before beginning the experiment.

Remember that the team will decide what volume of water to use in the calorimeter and what container to use as a calorimeter.

1. Begin heating a sample of water in the 400-mL beaker. Use enough water to completely submerge the metal sample.
2. Determine the mass of the sample of metal. Record the mass and identity of the metal.
3. Using tongs, gently lower the metal into the beaker of water being heated. Continue heating the metal and water until the water begins to boil.
4. Pour the predetermined volume of tap water into the calorimeter. Record the volume and temperature of the tap water.
5. When the water in the beaker begins to boil, record the temperature of the boiling water.
6. Turn off the hot plate or Bunsen burner. Hold the calorimeter in one hand near the beaker of hot water. Using tongs, carefully remove the metal from the boiling water and immediately place it in the calorimeter.
7. Monitor the temperature of the water in the calorimeter. Record the maximum temperature of the water.
8. Repeat procedure if necessary.

SAFETY INFORMATION

1. Be careful when placing the metal in and removing the metal from the hot water.
2. Follow all safety procedures for use of a hot plate or Bunsen burner.
3. After reviewing your procedure, the instructor will discuss any safety precautions that are specific to your experiment.

TABLE OF ACCEPTED VALUES

Compare the value that you calculated for specific heat with the following table of accepted values.

METAL	SPECIFIC HEAT (cal/g · °C)	SPECIFIC HEAT (J/g · °C)
Aluminum	2.15×10^{1}	8.99×10^{-1}
Copper	9.2×10^{-2}	3.8×10^{-1}
Lead	3.8×10^{-2}	1.6×10^{-1}
Nickel	1.06×10^{-1}	4.43×10^{-1}
Tin	5.1×10^{-2}	2.1×10^{-1}
Zinc	9.28×10^{-2}	3.88×10^{-1}
Cadmium	6.0×10^{-2}	2.5×10^{-1}
Brass	9.0×10^{-2}	3.8×10^{-1}
Iron or steel	1.1×10^{-1}	0.46

QUESTIONS FOR FURTHER THOUGHT

1. Calculate the quantity of heat, in calories, that would be required to heat 30.0 g of aluminum from 25.0°C to 40.0°C. (NOTE: the specific heat of aluminum is given in the chart.)
2. Substance *x* has a mass of 20.0 g and is found to absorb 800 calories as it is heated from 22.0°C to 80.0°C. What is the specific heat of the metal?
3. A metal is found to absorb 1000 calories as it is heated from 40.0°C to 60.0°C. How much heat would the same sample absorb if it were heated from 60.0°C to 80.0°C?

TEACHER'S NOTES

A lab procedure is included with this experiment due to the length of time that might be required for students to develop their own experiment. If you prefer to have students design their own experiment, simply propose the problem to the class or skip to experiment 10, Comparing the specific heat of metals and nonmetals.

The equations to be used in the experiment are not included in the background section of the lab. It is assumed that the students have already completed an experiment involving the concepts and equations of calorimetry. It is left to your discretion whether the students should be provided with the needed formulas.

OPTIONS

1. Even if students are following the procedure provided, the students can perform their own variations (e.g., changing the volume of water in the calorimeter, using several samples of the same metal, using samples of different metals, or varying the container used as the calorimeter). Different teams can do different variations of the given procedure and then compare and discuss their methods and results.
2. Students could repeat the procedure, only this time place the metal in a sample of ice water (rather than boiling water) before transferring it to the calorimeter. Students could calculate and compare the value for specific heat obtained using each method. Alternatively, some teams could use the boiling water and some teams could use the ice water and then compare their results.
3. Have the students do this experiment, following the provided procedure, and then have them complete experiment 10, Comparing the specific heat of metals and nonmetals, which is entirely designed by the students.
4. You can identify the metal and have the students calculate their percent error at the conclusion of the lab or have the students try to identify the metal from the table of accepted values.

NOTES

1. Indicate to the students the preferred unit for specific heat.
2. The metal samples used in this experiment should be large enough that they can be removed from the boiling water with tongs. Avoid using metal shot. The physics department is a good source of these metals if they aren't available in the chemistry department. If the metals are the type that have hooks attached at the top, students can tie a long piece of a strong thread to the hook and use the thread to lower the metal into the hot water and later transfer it to the calorimeter.
3. Another option for the metals are the 1″-long zinc rectangles that are sold for use in a Kipp generator or the metals sold as density sets or specific

gravity sets. Emphasize to the students that they must be careful when adding or removing the metal from the beaker.

4. If students keep the water in the beaker near the boiling point, it is easy to perform several trials in the same lab period.

5. When choosing the amount of water that the students will pour into the calorimeter, they should be aware that the metal must be totally submerged so it doesn't release heat to the air, and that the smaller the volume of water, the greater the value for ΔT and therefore the easier it is to measure.

TIME

The amount of time needed for students to design and execute their own experiment can vary depending on the level and creativity of the class, their familiarity with designing their own investigations, and the complexity of the procedure they develop. The times below are based on the students following the procedure provided.

Average Time for This Experiment

Time needed for reviewing procedure with students: 15–20 minutes
Time needed to run the experiment: 30 minutes

TEAMS

This experiment is best performed by individual students or teams of two.

MATERIALS (PER TEAM)

Require the use of goggles, gloves, aprons/lab coats. The following supplies should be readily available for the students:

- Variety of containers that can serve as a calorimeter including polystyrene cups, beakers, and metal cups. (NOTE: the containers must be large enough that the metals can be completely submerged in the water)
- Metal samples
- 400-mL beaker
- 100-mL graduated cylinder
- Thermometer
- Tongs
- Heat source
- Ice (if students will be doing option #2 above)
- Strong thread (if students will be using the thread to raise and lower the metal into the water)
- Students may also ask for materials such as aluminum foil, plastic wrap, wax paper, and so on, for covering their cups to limit the amount of heat loss.

SAFETY PRECAUTIONS

Refer to the safety information given with the experiment. Material Safety Data Sheets for each substance used should be reviewed and made available to the students.

Remind the students to be careful when adding or removing the metal from the boiling water.

COMMON MISCONCEPTIONS, PROCEDURAL AND CALCULATION ERRORS

1. Students may use the mass of the metal when initially calculating ΔH. They need to understand that they are calculating the change in heat content (ΔH) of the water and therefore all three variables in the equation (ΔT, mass, and specific heat) must refer to the water in the calorimeter. Then, they are to assume that the change in heat content of the water is numerically the same as the change in heat content of the metal.

2. Students will often think that the metal and water have the same change in temperature (ΔT). The metal's change in temperature is the difference between the temperature of the boiling water and the final temperature in the calorimeter, while the water's change in temperature is the difference between the initial and final temperature of the water in the calorimeter.

3. Once students calculate the change in heat content for the water, and therefore, the metal, they may believe they have found the specific heat of the metal. Remind the students that specific heat is the quantity of heat lost by the object per unit of mass and per degree. The unit of their answer should be the unit specified.

LAB REPORT

There are several alternatives to writing a formal lab report given in the introduction of the book.

If students are writing a formal lab report it should include:

I. Title
II. Purpose
III. Data
IV. Calculations
V. Conclusion (including a discussion of sources of error)
VI. Percent error
VII. Answers to Questions (optional)

ANSWERS TO THE QUESTIONS

1. $\Delta H = m \times \Delta T \times c$
$= (30.0\text{ g})(15.0°\text{C})(0.215\text{ cal/g·°C})$
$= 96.8\text{ cal}$

2. $\Delta H = m \times \Delta T \times c$
$(800\text{ cal}) = (20.0\text{ g})(58.0°\text{C})(c)$
$0.690\text{ cal/g·°C} = \text{c}$

3. The metal would absorb 1000 calories if it were heated from 60.0°C to 80.0°C because it is still undergoing a 20°C temperature change.

9 Determining the Specific Heat of a Metal

SAMPLE LAB REPORT

PURPOSE

To determine the specific heat of lead.

DATA

Volume of water in cup = 100.0 mL (100.0 g)
Mass of metal = 29.01 g
Temperature of the boiling water and initial temperature of the metal = 99.8°C
Initial temperature of water in Styrofoam cup = 22.0°C
Final temperature of water in Styrofoam cup and metal = 23.0°C

SAMPLE CALCULATIONS

ΔT for the metal = 99.8°C − 23.0°C = 76.8°C
ΔT for the water in cup = 23.0°C − 22.0°C = 1.0 °C
Quantity of heat absorbed by the water:

$$\begin{aligned}(\Delta H) &= \text{mass}_{(\text{water})} \times \Delta T_{(\text{water})} \times c_{(\text{water})} \\ &= 100.0\text{ g} \times 1.0^\circ\text{C} \times 4.18\text{ J/g}\cdot{}^\circ\text{C} \\ &= 418\text{ J} = 420\text{ J}\end{aligned}$$

Quantity of heat released by the metal = 420 J
Specific heat of the metal = 420 J/(29.01 g)(76.8°C) = 1.9×10^{-1} J/g·°C

CONCLUSION

Based on our data, the specific heat of lead is 1.9×10^{-1} J/g·°C. The accepted value is 1.588×10^{-1} J/g·°C.

SOURCES OF ERROR

The following is a compilation of the sources of error that students might suggest:

1. Heat was transferred from the metal to the tongs.
2. Heat was transferred from the water in the Styrofoam cup to the environment.
3. Heat was transferred from the metal to the air when the heated metal was moved to the calorimeter.
4. Using two different thermometers (one in the boiling water and one in the cup) that may have different calibrations.
5. If the metal has a hook attached, the hook may be composed of a different metal.
6. The purity of the metal is unknown.
7. We can't be sure that the metal reached the temperature of the boiling water.

8. The hot water that was on the metal was also transferred to the cup and released heat into the cooler water.
9. Water left in the graduated cylinder means that the cup did not contain precisely 100.0 mL.
10. Human and mechanical error.

10 Comparing the Specific Heat of Metals and Nonmetals

THE EXPERIMENT

PURPOSE

To compare the specific heat of metals and nonmetals.

BACKGROUND

The specific heat (or specific heat capacity) of a material is the quantity of heat energy needed to raise a given mass of a material, 1°C. The unit for specific heat is J/g·°C (although the unit cal/g·°C is also acceptable).

In this experiment you will be given samples of different materials. Design an experiment to determine the specific heat of the materials, comparing the values you obtain for the metals to those calculated for the nonmetals.

PROCEDURE

NOTE: Consult safety information and obtain teacher approval before beginning the experiment. Design an experiment to determine the specific heat of the materials, comparing the values you obtain for the metals to those calculated for the nonmetals.

SAFETY INFORMATION

After reviewing your procedure, the instructor will discuss any safety precautions that are specific to your experiment.

QUESTIONS FOR FURTHER THOUGHT

1. Substance *x* has a specific heat of 2.0×10^{-3} J/g·°C while substance *y* has a specific heat of 8.0×10^{-3} J/g·°C. If samples of *x* and *y*, with equal mass, both absorb 1500 calories, which substance will undergo the greater change in temperature?
2. When choosing a material to serve as an insulator, should a material with a high specific heat or one with a low specific heat be chosen?

10 Comparing the Specific Heat of Metals and Nonmetals

TEACHER'S NOTES

In this experiment, students are asked to compare the specific heat of metals and nonmetals. This lab can be performed by students after completing experiment 9, Determining the specific heat of a metal (which includes directions to introduce students to the general procedure) or be completed independently.

OPTIONS

1. Students can complete this experiment without having completed experiment 9 if you do not want the students to have a procedure to follow.
2. You can combine experiments 9 and 10 by providing students with the general procedure given in experiment 9 and having students experiment with metals and nonmetals.
3. If students completed experiment 9, they can simply use nonmetals in this experiment and compare their specific heat values to those obtained for the metals in the previous lab.

NOTES

1. When choosing nonmetals for this experiment, select objects that can be easily removed from boiling (or ice) water with tongs, will not shatter when transferred from boiling water to cool water, and can be submerged in a Styrofoam cup. Options include:

   Rubber balls or golf balls,
   Painted dominoes or blocks of wood (to reduce the absorption of water),
   Rubber stoppers,
   Rocks,
   Small plastic figurines and other plastic toys.

2. For notes on the metals that can be used, see experiment 9, notes 2 and 3.
3. Specify which unit to use for specific heat.

TIME

The amount of time needed for students to design and execute their own experiment can vary depending on the level and creativity of the class, their familiarity with designing their own investigations, and the complexity of the procedure they develop.

Average Time for This Experiment

Time needed to design the experiment: 15 minutes (if familiar with calorimetry)
Time needed to run the experiment: 30 minutes

TEAMS

Teams of two work well.

MATERIALS (PER TEAM)

Require the use of goggles, gloves, and aprons/lab coats. The following supplies should be readily available for the students:

- Metals and nonmetals (see notes 1 and 2 above)
- Styrofoam cups (8 or 16 oz)
- Thermometers (2)
- 250–400 mL beakers
- Heat source (hot plate or Bunsen burner)
- Ice (optional)
- Strong thread
- Tongs
- 100-mL graduated cylinder
- Balance

Students may ask for materials to cover their calorimeter to reduce the loss of heat to the air. Examples include covers for the cups, aluminum foil, plastic wrap, and so on.

SAMPLE PROCEDURES

If students completed experiment 9, their procedure will probably be similar to the one given with the experiment. One variation that students may perform is placing the object (metal or nonmetal) in ice water before moving it to the calorimeter. This will produce a smaller ΔT and therefore may not be as accurate.

SAFETY PRECAUTIONS

Discuss with the students any safety precautions that are specific to the procedure they develop. Material Safety Data Sheets for each substance used should be reviewed and made available to the students.

Students should follow all safety procedures when using a hot plate or Bunsen burner.

COMMON MISCONCEPTIONS, PROCEDURAL AND CALCULATION ERRORS

1. Students might forget to obtain the mass of the metal before beginning the experiment.
2. Students might use the mass of the metal, rather than the mass of the water in the cup, when calculating ΔH.
3. Students may confuse the change in temperature of the object with the change in temperature of the water in the calorimeter.
4. Students might calculate ΔH correctly, but leave this value as the specific heat of the object rather than converting to the unit J/g·°C.

LAB REPORT

There are several alternatives to writing a formal lab report given in the introduction. If students are writing their own lab report it should include:

I. Title
II. Purpose
III. Materials
IV. Procedure
V. Data
VI. Calculations
VII. Conclusion (including a discussion of sources of error)
VIII. Answers to Questions (optional)

ANSWERS TO THE QUESTIONS

1. Because $\Delta H = (m)(\Delta T)(c)$, ΔT and c (specific heat) are inversely proportional when ΔH and mass are held constant. Because substance *y* has a greater specific heat than substance *x*, substance *y* will undergo a smaller change in temperature. More specifically, because *y* has four times the specific heat of *x*, it will undergo ¼ the change in temperature. (Students might calculate the actual change in temperature for *x* and *y*, choosing a value for the mass of the samples.)
2. When choosing an insulator a substance with a high specific heat should be chosen because having a high specific heat means it will be able to absorb a large quantity of heat and undergo a smaller change in temperature than a substance with a lower specific heat.

10 Comparing the Specific Heat of Metals and Nonmetals

SAMPLE LAB REPORT

NOTE: Students in this experiment are only working with nonmetals and then comparing these values to those they obtained in experiment 9.

PURPOSE

To compare the specific heat of metals and nonmetals.

MATERIALS

Safety equipment: goggles, gloves and apron
Small rock
Plastic figurine
Golf ball
Rubber stopper
Styrofoam cup
Thermometer (2)
100-mL graduated cylinder
400-mL beaker
Tongs
Hot plate
Balance

PROCEDURE

1. Fill the 400-mL beaker about 3/4 full with tap water.
2. Begin heating the beaker on the hot plate.
3. Determine the mass of each of the four objects to be studied.
4. Add the rock and a thermometer to the hot water.
5. Pour 100.0 mL of tap water into the Styrofoam cup and record the temperature of the water.
6. When the heated water begins to boil, wait 4 minutes for the rock and water to reach the same temperature and record the temperature of the boiling water.
7. Using the tongs, remove the rock from the hot water and immediately place it in the calorimeter.
8. Monitor the temperature of the water in the calorimeter until it reaches a maximum value.
9. Record the maximum temperature of the water.
10. Repeat steps 4–9 with the other objects.

DATA

	ROCK	PLASTIC FIGURINE	RUBBER STOPPER	GOLF BALL
Mass (g)	75.27	26.49	28.44	45.34
Temp. of boiling water (°C)	99.1	100.6	100.2	100.1
Initial temp. of water in calorimeter (°C)	21.3	22.1	22.8	21.7
Final temp. of water in calorimeter (°C)	32.2	25.3	25.9	26.8
Change in temp. of water in calorimeter (°C)	10.9	25.3	3.1	5.1
Change in temperature of object (°C)	66.9	75.3	74.3	73.3

CALCULATIONS

Rock

$$\Delta H_{water} = \text{mass of water} \times \Delta T \text{ of water} \times \text{specific heat of water}$$
$$= 100.0 \text{ g} \times 10.9°\text{C} \times 1.00 \text{ cal/g·°C}$$
$$= 1.09 \times 10^3 \text{ cal}$$
$$\Delta H_{rock} = 1.09 \times 10^3 \text{ cal}$$

$$\text{Specific heat of the rock} = \frac{1.09 \times 10^3 \text{ cal}}{(75.27 \text{ g})(66.9°\text{C})} = 0.216 \text{ cal/g} \cdot °\text{C}$$

Plastic Figurine

$$\Delta H_{water} = \text{mass of water} \times \Delta T \text{ of water} \times \text{specific heat of water}$$
$$= 100.0 \text{ g} \times 3.2°\text{C} \times 1.00 \text{ cal/g·°C}$$
$$= 3.2 \times 10^2 \text{ cal}$$
$$\Delta H_{figurine} = 3.2 \times 10^2 \text{ cal}$$

$$\text{Specific heat of the plastic figurine} = \frac{3.2 \times 10^2 \text{ cal}}{(26.49 \text{ g})(75.3°\text{C})} = 0.16 \text{ cal/g} \cdot °\text{C}$$

Rubber Stopper

$$\Delta H_{water} = \text{mass of water} \times \Delta T \text{ of water} \times \text{specific heat of water}$$
$$= 100.0 \text{ g} \times 3.1°\text{C} \times 1.00 \text{ cal/g·°C}$$
$$= 3.1 \times 10^2 \text{ cal}$$
$$\Delta \text{H}_{rubber\ stopper} = 3.1 \times 10^2 \text{ cal}$$

$$\text{Specific heat of the stopper} = \frac{3.1 \times 10^2 \text{ cal}}{(28.44 \text{ g})(74.3°\text{C})} = 0.15 \text{ cal/g} \cdot °\text{C}$$

Golf Ball

$$\Delta H_{water} = \text{mass of water} \times \Delta T \text{ of water} \times \text{specific heat of water}$$
$$= 100.0 \text{ g} \times 5.1°\text{C} \times 1.00 \text{ cal/g·°C}$$
$$= 5.1 \times 10^2 \text{ cal}$$
$$\Delta H_{golf\ ball} = 5.1 \times 10^2 \text{ cal}$$

$$\text{Specific heat of the golf ball} = \frac{5.1 \times 10^2 \text{ cal}}{(45.34 \text{ g})(73.3°\text{C})} = 0.15 \text{ cal/g} \cdot °\text{C}$$

CONCLUSION

The specific heat values of the nonmetals were:

Rock = 0.216 cal/g·°C
Plastic figurine = 0.16 cal/g·°C
Rubber stopper = 0.15 cal/g·°C
Golf ball = 0.15 cal/g·°C

These values seem to be between those for most of the metals studied in our previous experiment (#9) and the specific heat of water (1.00 cal/g·°C).

SOURCES OF ERROR

The following is a compilation of sources of error that students may suggest:

1. The tongs may have absorbed some heat from the object.
2. Loss of heat from the water in the calorimeter to the environment.
3. Loss of heat from the object to the air during the transfer.
4. Using two different thermometers (one in the boiling water and one in the calorimeter) that may have different calibrations.
5. We can't be sure that the object reached the temperature of the boiling water.
6. Hot water that was on the object also transferred heat to the cup.
7. Water was left in the graduated cylinder so the cup did not contain precisely 100.0 mL.
8. Human and mechanical error.

11 Determining the Mass of a Sample of Metal Using Its Specific Heat

THE EXPERIMENT

PURPOSE

To determine the mass of a sample of metal using its specific heat.

BACKGROUND

The specific heat (or specific heat capacity) of a material is the quantity of heat energy needed to raise a given mass of a material 1°C. The unit of specific heat is J/g·°C (although the unit cal/g·°C is also acceptable).

In this experiment you will be given a sample of metal. Design an experiment to determine the mass of the metal using its known specific heat.

PROCEDURE

NOTE: Consult safety information and obtain teacher approval before beginning the experiment. Design an experiment to determine the mass of a sample of metal using its known specific heat.

SAFETY INFORMATION

After reviewing your procedure, the instructor will discuss any safety precautions that are specific to your experiment.

ACCEPTED VALUES OF SPECIFIC HEAT FOR VARIOUS METALS

METAL	SPECIFIC HEAT (cal/g · °C)	SPECIFIC HEAT (J/g · °C)
Aluminum	2.15×10^{-1}	8.99×10^{-1}
Copper	9.2×10^{-2}	3.8×10^{-1}
Lead	3.8×10^{-2}	1.6×10^{-1}
Nickel	1.06×10^{-1}	4.43×10^{-1}
Tin	5.1×10^{-2}	2.1×10^{-1}
Zinc	9.28×10^{-2}	3.88×10^{-1}
Cadmium	6.0×10^{-2}	2.5×10^{-1}

(*chart continues*)

METAL	SPECIFIC HEAT (cal/g · °C)	SPECIFIC HEAT (J/g · °C)
Brass	9.0×10^{-2}	3.8×10^{-1}
Iron or steel	1.1×10^{-1}	4.6×10^{-1}

QUESTIONS FOR FURTHER THOUGHT

1. A sample of aluminum is found to absorb 430 calories as it is heated from 60°C to 80°C. What is the mass of this sample of aluminum?
2. A sample of water is found to release 800 calories as it is cooled from 50°C to 20°C. What is the mass of this sample of water?

11 Determining the Mass of a Sample of Metal Using Its Specific Heat

TEACHER'S NOTES

In this experiment, students are asked to determine the mass of a sample of metal using its accepted specific heat. This lab should be performed after students have completed experiment 9, Determining the specific heat of a metal or experiment 10, Comparing the specific heat of metals and nonmetals. Experiment 9 includes a procedure and is intended to introduce students to the general procedures used in calorimetry. After completing that experiment, students should be able to design their own procedure for this experiment and should attempt to reduce any potential sources of error that they identifed during experiment 9.

OPTIONS

1. You may want to determine the mass of each sample of metal before distributing the samples to the students.
2. You can combine experiments 10 and 11 by dividing the class in half and having each half perform one of the experiments.

NOTES

1. For notes on the metals that can be used see experiment 9, notes 2 and 3.
2. You should identify the metal to the students.
3. You may want to put classroom balances away so students can't use them during the experiment.

TIME

The amount of time needed for students to design and execute their own experiment can vary depending on the level and creativity of the class, their familiarity with designing their own investigations, and the complexity of the procedure they develop.

Average Time for This Experiment

Time needed to design the experiment: 15 minutes (if students completed experiment 9)
Time needed to run the experiment: 30–45 minutes

TEAMS

Teams of two work well.

MATERIALS (PER TEAM)

Require the use of goggles, gloves, and aprons/lab coats. The following supplies should be readily available for the students:

Metal samples
Styrofoam cups (8 or 16 ounce)
Thermometers
250–400 mL beakers
Heat source (hot plate or Bunsen burner)
Ice (optional)
Strong thread
Tongs
100-mL graduated cylinder

Students may ask for materials to cover their calorimeter to reduce the loss of heat to the air. Examples include covers for the cups, aluminum foil, plastic wrap, and so on.

SAMPLE PROCEDURES

If students completed experiment 9, their procedure will probably be similar to the one given with the experiment. One variation that students may perform is placing the metal in ice water before moving it to the calorimeter. This will produce a smaller ΔT and therefore may not be as accurate.

SAFETY PRECAUTIONS

Discuss with the students any safety precautions that are specific to the procedure they develop. Material Safety Data Sheets for each substance used should be reviewed and made available to the students.

Students should follow all safety procedures when using a hot plate or Bunsen burner.

COMMON MISCONCEPTIONS, PROCEDURAL AND CALCULATION ERRORS

Students may confuse the change in temperature of the object with the change in temperature of the water in the calorimeter.

LAB REPORT

There are several alternatives to writing a formal lab report given in the introduction. If students are writing their own lab report it should include:

I. Title
II. Purpose
III. Materials
IV. Procedure
V. Data
VI. Calculations
VII. Conclusion (including a discussion of sources of error)
VIII. Answers to Questions (optional)

ANSWERS TO THE QUESTIONS

1. For the sample of metal:

   ΔH = (mass)(ΔT)(specific heat)
   430 calories = (mass)(20 °C)(0.215 cal/g°·C)
   Mass of aluminum = 100 g

2. For the sample of water:

   ΔH = (mass)(ΔT)(specific heat)
   800 calories = (mass)(30°C)(1.00 cal/g·°C)
   Mass of water = 27 g

11 Determining the Mass of a Sample of Metal Using Its Specific Heat

SAMPLE LAB REPORT

PURPOSE

To determine the mass of a sample of copper using its specific heat.

MATERIALS

Copper cylinder
Styrofoam cup
Thermometer
Hot plate
Tongs
100-mL graduated cylinder
400-mL beaker

PROCEDURE

1. Pour approximately 300 mL of water into the 400-mL beaker.
2. Begin heating the water on the hot plate.
3. Carefully place the copper cylinder in the water.
4. Place a thermometer in the water and record the boiling point.
5. Pour 100.0 mL of tap water into the Styrofoam cup. Record the temperature of the water.
6. After the metal has been in the boiling water for a few minutes, carefully remove it from the water and place it in the Styrofoam cup.
7. Monitor the temperature of the water in the Styrofoam cup, recording the maximum temperature.

DATA

Volume of water in cup = 100.0 mL (100.0 g)
Temperature of the boiling water and initial temperature of the metal = 103.0°C
Initial temperature of water in polystyrene cup = 19.0°C
Final temperature of water in polystyrene cup and metal = 23.5°C

CALCULATIONS

ΔT for the metal = 103.0°C − 23.5°C = 79.5°C
ΔT for the water in cup = 23.5°C − 19.0°C = 4.5°C

Quantity of heat absorbed by the water:

$$
\begin{aligned}
(\Delta H) &= \text{mass}_{(\text{water})} \times \Delta T_{(\text{water})} \times c_{(\text{water})} \\
&= 100.0\text{ g} \times 4.5°\text{C} \times 4.18\text{ J/g·°C} \\
&= 1881\text{ J} = 1.9 \times 10^3\text{ J}
\end{aligned}
$$

Quantity of heat released by the metal = 1.9×10^3 J
Mass of the metal = ΔH/(change in temperature)(specific heat)
= 1.9×10^3 J/(79.5°C)(0.38 J/g·°C) = 66 g

CONCLUSION

Based on our data, the mass of our sample of copper was 66 g. Our instructor told us that the measured mass was 63.84 g. This would give us a percent error of 3.4.

SOURCES OF ERROR

The following is a compilation of the sources of error that students might suggest:

1. Transfer of heat from the metal to the tongs.
2. Transfer of heat from the water in the Styrofoam cup to the environment.
3. Transfer of heat from the metal to the air during the transfer.
4. Using two different thermometers (one in the boiling water and one in the polystyrene cup) that may have different calibrations.
5. If the metal has a hook attached, the hook may be composed of a different metal.
6. The purity of the metal.
7. We can't be sure that the metal reached the temperature of the boiling water.
8. Hot water that was on the metal was also transferred to the cup and released heat into the cooler water.
9. Water left in the graduated cylinder means that the cup did not contain precisely 100.0 mL.
10. Human and mechanical error.

12 Celsius versus Fahrenheit

THE EXPERIMENT

PURPOSE

To find a mathematical relationship between the Celsius and Fahrenheit temperature scales.

BACKGROUND

Three scales often used to measure temperature are Celsius, Fahrenheit, and Kelvin. The Fahrenheit scale was developed by Gabriel Fahrenheit (1686–1736). He devised a scale in which 32°F represented the melting point of water and 212°F represented the boiling point (both at sea level). The Celsius scale was invented by the Swedish astronomer Anders Celsius (1701–1744). On the Celsius scale, the melting point of water was chosen to be 0°C and the boiling point 100°C (at standard pressure). The SI unit of temperature is the Kelvin scale, named after the English physicist and mathematician William Thomson, Lord Kelvin (1824–1907). The Kelvin scale is called the absolute scale because 0 K (notice the degree symbol is not used with the Kelvin scale) represents absolute zero, the coldest possible temperature.

It is often necessary in the lab to convert from one unit of temperature to another. To convert between the two temperature scales, an equation that relates the two scales is used. To convert between Celsius and Kelvin, the equation is $K = °C + 273$.

In this experiment you will be given Celsius and Fahrenheit thermometers. Use the thermometers to determine a mathematical relationship between the two temperature scales.

PROCEDURE

NOTE: Consult safety information and obtain teacher approval before beginning the experiment. Use the Celsius and Fahrenheit thermometers water to determine a mathematical relationship between the Celsius and Fahrenheit temperature scales.

SAFETY INFORMATION

After reviewing your procedure, the instructor will discuss any safety precautions that are specific to your experiment.

If a thermometer breaks, notify the instructor immediately.

QUESTIONS FOR FURTHER THOUGHT

1. Define temperature.
2. Define heat.
3. Has absolute zero ever been reached? What happens at absolute zero?
4. Room temperature is about 20°C. What is this temperature in Fahrenheit? Kelvin?

TEACHER'S NOTES

Students are often aware of the formulas used to convert between the Celsius and Fahrenheit temperature scales:

$$°F = (\frac{9}{5})\, °C + 32$$

$$°C = (\frac{5}{9})(°F - 32)$$

but do not understand how these (and other) formulas are derived. This experiment shows that the relationship given by a graph produces an equation that mathematically relates two variables.

This lab can be performed when students are studying temperature scales, or as an introduction to a unit, such as the gas laws, that involves graphical relationships.

OPTIONS

This lab can also be performed with temperature probes and a computer interface or a Calculator-Based Laboratory (CBL™). Simply calibrate one of the probes to read in the Celsius scale and the other probe to read in Fahrenheit.

NOTES

1. Students need to be familiar with determining the equation of a best-fit line. They also need to know the equation of a line: $y = mx + b$. You might want to make sure that the students understand these concepts before beginning the experiment.
2. If the students have access to a computer and a graphing program or a graphing calculator, the students can enter their data and use the program to obtain the equation.
3. Students should come close to obtaining the relationship given by the first or second equation. The equation they obtain will depend on which temperature scale they set as the x- and y-axis. The fraction $9/5 = 1.80$ and $5/9 = 0.556$. Students should obtain approximately these values for the slope of their line.
4. Fahrenheit thermometers can be purchased through many chemical supply companies.

TIME

The amount of time needed for students to design and execute their own experiment can vary depending on the level and creativity of the class, their familiarity with designing their own investigations, and the complexity of the procedure they develop.

Average Time for This Experiment

Time needed to design the experiment: 15 minutes
Time needed to run the experiment: 30 minutes

TEAMS

Students may perform the experiment individually or work in teams of two.

MATERIALS (PER TEAM)

Require the use of goggles, gloves, and aprons/lab coats. The following supplies should be readily available for the students:

- Celsius and Fahrenheit thermometers (or two temperature probes with a computer interface or CBL™)
- Beaker
- Heat source
- Ice
- Insulated gloves or beaker tongs
- Graphing program (optional)

SAMPLE PROCEDURES

1. Some students may want to simply plot the known values for the freezing point and boiling point of water and determine the equation of the best-fit line using these two points without having to obtain any data. You can decide if this is adequate.
2. Students will probably place the two thermometers in water at various temperatures and record the temperature of the water in both the Celsius and Fahrenheit scales. Then they will graph the data to produce the equation of the best-fit line.

SAFETY PRECAUTIONS

Discuss with the students any safety precautions that are specific to the procedure they develop. Material Safety Data Sheets for each substance used should be reviewed and made available to the students.

Students should follow all safety precautions when using a hot plate or Bunsen burner.

COMMON MISCONCEPTIONS, PROCEDURAL AND CALCULATION ERRORS

1. Students might take the temperature of one sample of water and believe they are finished. This will only produce one point on the graph unless they add two known points such as the melting and boiling points of water.
2. Students might use the thermometers in different beakers rather than using them to obtain the temperature of a given sample of water.
3. Some students might have difficulty obtaining the equation of the best-fit line.
4. Some students may not know how to analyze the data to determine a mathematical relationship between the variables. You may want to suggest graphing the data and letting them take it from there.

LAB REPORT

There are several alternatives to writing a formal lab report given in the introduction. If students are writing their own lab report it should include:

I. Title
II. Purpose
III. Materials
IV. Procedure
V. Data
VI. Graph
VII. Conclusion (including a discussion of sources of error)
VIII. Answers to Questions (optional)

ANSWERS TO THE QUESTIONS

1. Temperature is a measure of the average kinetic energy of the particles in a substance. It reflects the random motion of the particles.
2. Heat is the energy that is transferred between two objects of different temperatures.
3. Absolute zero has never been reached. At absolute zero, all molecular motion would stop.
4. Room temperature is about 293 K or 68°F.

12 Celsius versus Fahrenheit

SAMPLE LAB REPORT

PURPOSE

To find a mathematical relationship between the Celsius and Fahrenheit temperature scales.

MATERIALS

Safety equipment: goggles, gloves, and apron
Celsius and Fahrenheit thermometers
250-mL beaker
Hot plate
Ice

PROCEDURE

1. Pour approximately 150 mL of tap water into the 250-mL beaker.
2. Place both the Celsius and Fahrenheit thermometers in the water.
3. Record the temperature of the water using both thermometers.
4. Place the beaker on the hot plate and begin heating.
5. After a few minutes, record the temperature of the water using both thermometers.

6. Repeat step 5 after a few more minutes.
7. Pour the water in the sink and add another 150-mL sample of tap water.
8. Add a few ice cubes to the water.
9. Using the thermometers, gently stir the ice with the thermometers. After a few minutes, record the temperature of the ice water mixture with both thermometers.

DATA

CELSIUS	FAHRENHEIT
23.3°	72.1°
31.9°	87.1°
2.5°	35.3°
101.4°	209.9°

GRAPH

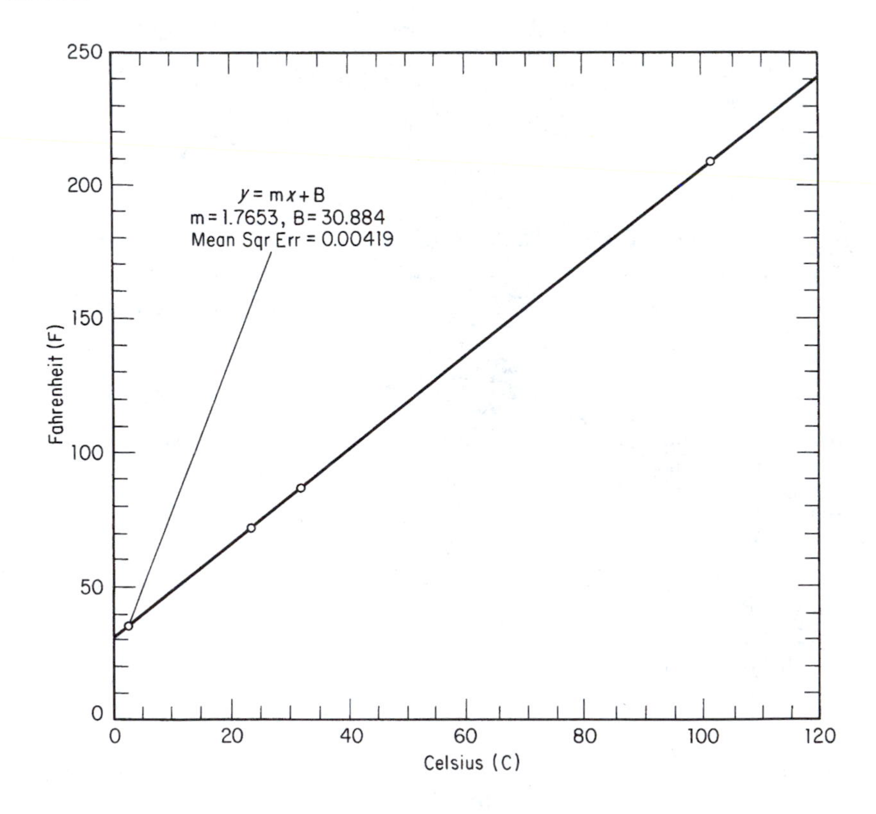

CONCLUSION

The relationship between Celsius and Fahrenheit is given as the equation of the best-fit line on the graph, which is: °F = 1.77 (°C) + 30.9

SOURCES OF ERROR

The following is a compilation of sources of error that students may suggest:

1. Different parts of the beaker may be at different temperatures due to uneven heating.
2. It was difficult to read the thermometers at exactly the same time.
3. The two thermometers probably aren't 100% accurate.
4. We didn't take the temperature of the water at the boiling point, which probably would have been better because the temperature would have been constant.

13 Name That Temperature

THE EXPERIMENT

PURPOSE

To determine the temperature of a sample of water (provided by the instructor) using a blank thermometer.

BACKGROUND

Scientific instruments often require calibration. Calibration is the preparation of a useable scale on a measuring instrument. To calibrate an instrument, standards (items of known value) are used to set the scale. For example, when calibrating an electronic balance, an object of known mass would be used. When calibrating a pH meter, two solutions of known pH would be used as standards to set the scale. Likewise, after a scale has been set, its accuracy can be checked with standards.

In this experiment, you will be calibrating a blank thermometer (one that doesn't have a scale). At a certain point in the lab period, the instructor will ask you to use your calibrated thermometer to measure the temperature of a given sample of water. The instructor will specify which unit or units of temperature may be used.

MATERIALS

(Check with the instructor for other needed supplies.)

- Safety equipment: goggles, gloves, and apron
- Blank thermometer
- Wax pencil
- Clear tape
- Beaker
- Ruler

PROCEDURE

NOTE: Consult safety information and obtain teacher approval before beginning the experiment. Design a procedure that will allow you to identify the temperature of a given sample of water using the blank thermometer provided.

SAFETY INFORMATION

1. Do not place the thermometer in your mouth.
2. The use of alcohol thermometers is recommended. If using mercury thermometers, review the Material Safety Data Sheet on mercury.
3. After reviewing your procedure, your instructor will discuss any safety precautions that are specific to your experiment.

QUESTIONS FOR FURTHER THOUGHT

1. Why is calibration important in the scientific laboratory?
2. Each of the following measuring instruments has a scale that must be calibrated. What is the function of each?

   A. Micrometer
   B. Altimeter
   C. Hydrometer
   D. Pyrometer
   E. Tachometer

3. Name some things found in the home that are calibrated.

13 Name That Temperature

TEACHER'S NOTES

Students are often unfamiliar with the concept of calibration. This is a good experiment to introduce the concept, particularly if students will be working with computer probes, CBLs™, spectrophotometers, or other instruments that require calibration during the year.

OPTIONS

1. The warm water should be in an insulated container if possible, so the temperature doesn't fluctuate too much while the different teams of students use their thermometers to measure its temperature. You may want to have all the teams place their thermometer in the water at the same time along with your thermometer so the temperature of the water is the same for everyone.
2. Inform the students that the warm water will be at your desk at the end of the class (or the beginning of the next lab period).

NOTES

1. Students can write directly on the thermometer using the wax pencil, or, if the students are going to use a marking pen, you can have the students cover the thermometer with clear tape and make their hash marks on the cellophane. (Use a pen that is not water soluble but be certain that the students are not writing directly on the glass unless they are using the wax pencil).
2. Blank spirit-filled thermometers are sold by Flinn Scientific.[1]
3. Students are not told that there will be ice or a heat source available for their use. However, you should have two or three trays of ice ready for each class as well as hot plates or bunsen burners. Students may also request a Styrofoam cup for their ice water.
4. Students may ask for a thermometer that is already calibrated, but that would defeat the purpose of the lab.
5. Students may ask to check a barometer and a reference book to determine the boiling point of water under the current atmospheric pressure.
6. If there are only enough thermometers available for one class, the entire experiment will have to be completed in one class period, so the hash marks and tape can be removed for the next class.

TIME

The amount of time needed for students to design and execute their own experiment can vary depending on the level and creativity of the class, their familiarity with designing their own investigations, and the complexity of the procedure they develop.

Average Time for This Experiment

Time needed to design the experiment: 20 minutes
Time needed to run the experiment: 30–40 minutes

TEAMS

Teams of two work well.

MATERIALS (PER TEAM)

Require the use of goggles, gloves, and aprons/lab coats. The following supplies should be readily available for the students:

- Blank thermometer
- Beaker (250–400 mL)
- Heat source
- Ice
- Styrofoam cup
- Wax pencil
- Clear tape
- Ruler

SAMPLE PROCEDURES

Most students will decide on a procedure in which they measure two different known temperatures using their thermometer and then create a scale based on two hash marks. The known temperatures students might want to use include:

A. Ice water
B. Boiling water
C. Room temperature (by reading a thermostat in the room)
D. Room temperature (by estimation)
E. Body temperature (by tucking the thermometer in their armpit)
F. Body temperature (by placing the thermometer in their mouth). This is not recommended.

Students may make marks directly on the thermometer with a wax pencil or a marking pen on tape. Other teams may make a scale on paper which they cut out and tape to the thermometer.

SAFETY PRECAUTIONS

Discuss with the students any safety precautions that are specific to the procedure they develop. Material Safety Data Sheets for each substance used should be reviewed and made available to the students.

COMMON MISCONCEPTIONS, PROCEDURAL AND CALCULATION ERRORS

Students may mark only one known point on the thermometer, such as the freezing point or boiling point of water.

LAB REPORT

(A lab report is not associated with this experiment.)

ANSWERS TO THE QUESTIONS

1. Calibration is important because it ensures the accuracy of scientific instruments.
2. A micrometer is used to measure very small distances. An altimeter is used to measure altitude; a hydrometer is used to measure the specific

gravity of a substance; a pyrometer is used to measure high temperatures, especially those of furnaces; and a tachometer is used to measure speed.
3. Objects in the home that are calibrated include thermostats, thermometers, refrigerators, ovens, bathroom scales, measuring cups, and rulers.

REFERENCES

1. Flinn Scientific, P.O. Box 219, Batavia, IL 60510-0219.

14 The Effect of Temperature on the Rate of a Clock Reaction

THE EXPERIMENT

PURPOSE

To study the effect of temperature on the rate of a clock reaction.

BACKGROUND

The rate of a reaction is the time required for a given quantity of reactants to be converted to product(s). The rate of a reaction is determined by many factors including temperature.

This experiment involves mixing two solutions (identified as solution 1 and solution 2). After the two solutions are combined, a marked change in the color of the solution will occur after a certain amount of time.

Design an experiment to study the relationship betweeen temperature and the rate of this reaction.

PROCEDURE

NOTE: Consult safety information and obtain teacher approval before beginning the experiment.

You will be given 50 mL of solution 1 and 50 mL of solution 2. Design an experiment to study the relationship betweeen temperature and the rate of reaction.

SAFETY INFORMATION

1. Keep the temperature of solutions 1 and 2 between 5°C and 75°C.
2. Heat and/or cool the solutions using a warm or cold water bath.
3. Dispose of the final solution as directed by your instructor.
4. After reviewing your procedure, the instructor will discuss any safety precautions that are specific to your experiment.

QUESTIONS FOR FURTHER THOUGHT

1. Explain the effect of temperature on the rate of the reaction in terms of both the kinetic and collision theories.

2. What are other factors that affect the rate of a reaction? Explain the effect of each of these variables in terms of the collision theory.

14 The Effect of Temperature on the Rate of a Clock Reaction

TEACHER'S NOTES

This is a popular lab among students due to the color change that occurs at the conclusion of the reaction.

A common rule of thumb regarding temperature and the rate of a reaction is that for every 10°C temperature increase, the rate of a reaction doubles. This rule only applies to certain reactions at moderate temperatures, but students may obtain data that support this trend.

Before beginning the experiment, you may want to clarify the difference between the rate of a reaction and the time of a reaction. Some students confuse the fact that an increase in the *rate* of a reaction is equivalent to a decrease in the *time* for the reaction.

The reaction that is taking place has the following mechanism[1]:

$$IO^-_{3(aq)} + 3HSO^-_{3(aq)} \rightarrow 3HSO^-_{4(aq)} + I^-_{(aq)}$$

$$6H^+_{(aq)} + IO^-_{3(aq)} + 5I^-_{(aq)} \rightarrow 3I_{2(aq)} + 3H_2O_{(l)}$$

$$I_{2(aq)} + starch_{(aq)} \rightarrow I_2{\cdot}[starch]_{(aq)}$$

OPTIONS

1. The experiment can be used when the students are studying the kinetic theory or the collision theory.
2. Students can use the same reaction to investigate the effect of concentration changes on the rate of a reaction.

NOTES

1. Students will often discover during the experiment that it is difficult to get solutions 1 and 2 to the same temperature even when they are in the same water bath. Students may decide to check the calibration of the two thermometers or average the temperature of 1 and 2.
2. It can save some time to have the students obtain all 50 mL of solution 1 and solution 2 at the beginning of the lab period. Then they can divide the solutions into smaller samples at their lab station.
3. When students analyze the data, they should see that as the temperature of the reaction increases, the rate of the reaction increases (or the time of the reaction decreases). This may be the extent of the analysis for some students, but other students may try to determine if a mathematical relationship exists between the variables.

If the students have already studied the gas laws and understand the relationship between Kelvin temperature and kinetic energy and how to determine both graphically and mathematically if variables are inversely proportional, they may convert their Celsius temperatures to Kelvin and graph the data or determine if the product of the two variables produces a constant.

Students should graph temperature (the dependent variable) on the x-axis and time of reaction (the independent variable) on the y-axis. Many students will extrapolate the graph so it crosses the x-axis. Have them consider whether or not they can assume that the relationship shown in the temperature range tested continues indefinitely in both directions.

TIME

The amount of time needed for students to design and execute their own experiment can vary depending on the level and creativity of the class, their familiarity with designing their own investigations, and the complexity of the procedure they develop.

Average Time for This Experiment

Time needed to design the experiment: 20 minutes
Time needed to run the experiment: 40–50 minutes

TEAMS

Teams of two to three work well.

MATERIALS (PER TEAM)

Require the use of goggles, gloves, and aprons/lab coats. The following materials should be readily available for the students:

- Solutions 1 and 2 (50 mL per team):
  - Solution 1: dissolve 4.3 g. of KIO_3 per liter of solution
  - Solution 2: combine 4.0 g. of soluble starch in 500 mL of water and heat (while stirring with a magnetic stirrer) until the solution is transparent. (If the solution is heated without stirring, the starch can burn.) After the solution has cooled, add 2.0 g. of $NaHSO_3$ and 5 mL of concentrated ulfuric acid. Dilute to 1.0 L. This solution should be fresh.
- Stopwatch or clock
- Test tubes
- 10-mL graduated cylinders (2)
- 50-mL graduated cylinders (2)
- Thermometers (2)
- Heat source (hot plate or Bunsen burner)
- Ice
- 250-mL or 400-mL beakers (to use as a hot or cold water bath)
- 100-mL beakers

SAMPLE PROCEDURES

1. Many students will probably design an experiment in which they mix a given volume of solution 1 with a given volume of solution 2 at room temperature and record the time required for the reaction. Then they will place two test tubes (one containing solution 1 and one containing solution 2) in a hot or cold water bath until they reach the desired temperature, at which point they mix the two solutions. The students will repeat the process at different temperatures.
2. Some students may decide to heat (or cool) solution 1 for three or four

trials and then heat (or cool) solution 2 for three or four trials to determine if one has a greater effect than the other.

SAFETY PRECAUTIONS

NOTE: Also refer to the safety information given with the experiment. Discuss with the students any safety precautions that are specific to the procedure they develop. Material Safety Data Sheets for each substance used should be reviewed and made available to the students.

Additional notes on sulfuric acid: Avoid contact with skin, eyes, and mucous membranes. Always add acid to water, never the reverse. Sulfuric acid is severely corrosive. Use and dispense in a fume hood. In the event of contact, wash affected parts with large amounts of water. Store in an acid cabinet. Use sand or vermiculite to absorb spills. If contact with eyes or skin occurs, rinse 15 minutes in a safety shower or eyewash and notify a physician.

COMMON MISCONCEPTIONS, PROCEDURAL AND CALCULATION ERRORS

1. Students might only use one thermometer, placing it in the warm or cold water bath and assume that the two solutions have reached the same temperature as the bath.
2. Students might place a thermometer in only solution 1 or only solution 2 and assume that the other solution is the same temperature.
3. Students may use only one thermometer to measure the temperature of both solution 1 and solution 2 while the solutions are in a bath and forget to rinse the thermometer before placing it in the other solution. This can cause a premature reaction.
4. Students may decide to place ice cubes directly in solution 1 or 2 to cool them and then remove the ice and combine the solutions. This will change the concentration of the solutions and introduce a new variable.

LAB REPORT

There are several alternatives to writing a formal lab report given in the introduction. If students are writing a formal lab report it should include:

I. Title
II. Purpose
III. Materials
IV. Procedure
V. Data
VI. Calculations and/or graphs
VII. Conclusion (including a discussion of sources of error)
VIII. Answers to Questions (optional)

ANSWERS TO THE QUESTIONS

1. As the temperature of each solution changes, the kinetic energy of the molecules also changes. For a collision to be effective and products to form, the reacting particles must collide at the correct angle and with enough energy to form an activated complex. At higher temperatures,

molecules collide more frequently, and a greater percentage collide with sufficient energy to form an activated complex; therefore, the rate of the reaction increases. The reverse is true as the temperature decreases.

2. Other factors that affect the rate of a reaction include the concentration of the reactants, the presence or absence of a catalyst, and the pressure (for reactions involving gases). An increase in the concentration of a reactant usually increases the rate of a reaction because there will be more collisions per unit of time. Catalysts speed up reactions by creating a mechanism that has a lower activation energy and therefore produce more effective collisions per unit of time. An increase in pressure in a reaction involving gaseous reactants will increase the rate of a reaction by producing more collisions per unit of time.

REFERENCES

1. *Chemistry: Connections to Our Changing World,* Lab Manual. Prentice Hall, Upper Saddle River, NJ, 1996, p. 353.

FOR FURTHER READING

Autuori, M.A., Brolo, A.G., Mateus, A.L.M.L. *Journal of Chemical Education.* 1989, 66, 852.

Brice, L.K. *Journal of Chemical Education.* 1980, 57, 152.

Lambert, J.L., Fina, G.T. *Journal of Chemical Education.* 1984, 61, 1037–1038.

14 The Effect of Temperature on the Rate of a Clock Reaction

SAMPLE LAB REPORT

PURPOSE

To study the effect of temperature on the rate of a clock reaction.

MATERIALS

- Safety equipment: goggles, apron and gloves
- Solutions 1 and 2
- Stopwatch
- 50-mL graduated cylinders (2)
- 10-mL graduated cylinders (2)
- Thermometers (2)
- Hot plate
- Ice
- Test tubes
- 100-mL beaker
- 250-mL beaker

PROCEDURE

1. Obtain 50.0 mL of solution 1 in one 50-mL cylinder and 50.0 mL of solution 2 in the other cylinder.
2. Pour 10.0 mL of solution 1 into a test tube and pour 10.0 mL of solution 2 into another test tube.

3. Measure the temperature of the two solutions.
4. Pour the two solutions into the 100-mL beaker and immediately begin timing the reaction with the stopwatch.
5. Record how long it takes for the solution to change color.
6. Dispose of the solution as directed by the instructor.
7. Pour another 10.0 mL of solution 1 into a clean test tube and 10.0 mL of solution 2 into another test tube.
8. Fill the 250-mL beaker until it is about 3/4 full.
9. Place a thermometer in each test tube.
10. Place both test tubes in the 250-mL beaker and begin heating until the solutions reach 35°C.
11. Immediately pour both solutions into the 100-mL beaker and time the reaction.
12. Repeat steps 6–10 at 45°C and 55°C.

DATA

TEMPERATURE (°C)	TEMPERATURE (K)	TIME (sec)
24	297	13.92
35	308	10.63
44	317	9.91
55	328	9.26

GRAPH

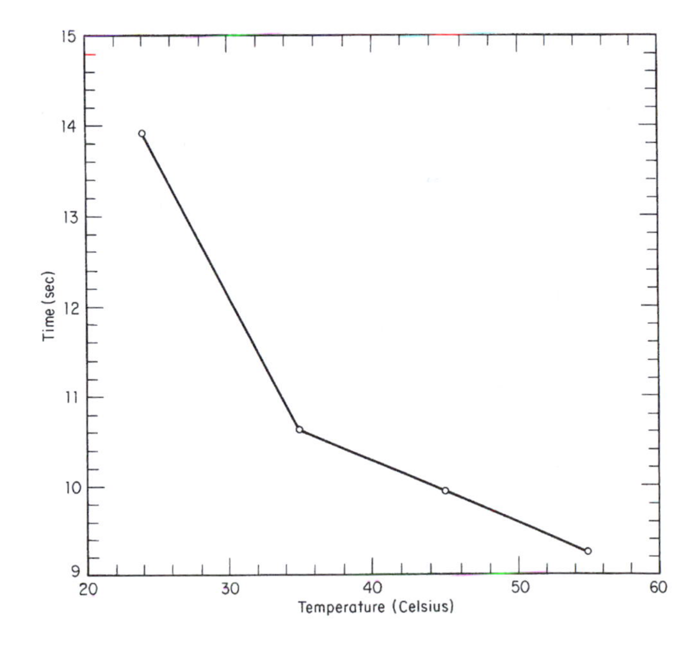

CALCULATIONS

1. Determine if Kelvin temperature and time are inversely proportional by calculating the product of the two variables to determine if it is a constant.

A. (297 K)(13.92 sec) = 4.13×10^3 K·sec
B. (308 K)(10.63 sec) = 3.27×10^3 K·sec
C. (317 K)(9.91 sec) = 3.14×10^3 K·sec
D. (328 K)(9.26 sec) = 3.04×10^3 K·sec

2. Determine if Celsius temperature and time are inversely proportional by calculating the product to determine if it is a constant.

A. (24°C)(13.92 sec) = 3.30×10^2°C·sec
B. (35°C)(10.63 sec) = 3.7×10^2°C·sec
C. (45°C)(9.91 sec) = 4.5×10^2°C·sec
D. (55°C)(9.26 sec) = 5.1×10^2°C·sec

CONCLUSION

The graph shows that the time of this reaction is inversely related to the Celsius temperature of the solutions, but because the product does not produce a constant (as shown in #2 above), we do not think that rate and Celsius temperature are inversely proportional.

The calculations do show that the time of the reaction may be inversely proportional to the Kelvin temperature. The product of (Kelvin temp.)(time)= is fairly close for trials B–D, but not for trial A. This could be due to the sources of error listed below. It is clear, however, that as the temperature of the reaction increased the rate of the reaction increased.

SOURCES OF ERROR

The following is a compilation of sources of error that students may suggest:

1. If the reaction is significantly endothermic or exothermic, the reaction itself could alter the temperature of the solutions.
2. For each trial the solutions were not always the same temperature.
3. Maintaining a constant temperature of the solutions is difficult.
4. As the solutions were heated, some evaporation could have occurred, which would alter the concentration of the reactants.
5. If water was still in the test tubes after rinsing, the concentration of the solutions could have been affected.
6. The use of two thermometers which probably had slightly different calibrations could have affected the data.

15 The Effect of Temperature on the Reaction Rate of Alka Seltzer™ in Water

THE EXPERIMENT

PURPOSE

To study the effect of temperature on the reaction rate of Alka Seltzer™ in water.

BACKGROUND

The rate of a reaction is the time required for a given quantity of reactants to be turned into product(s). The rate of a reaction is determined by many factors including the nature of the reactants, the concentration of the reactants, the temperature, the presence or absence of a catalyst, and the pressure (for reactions involving gases).

Alka Seltzer™ tablets contain heat-treated sodium bicarbonate, citric acid, and a salicylate analgesic. When the tablet is placed in water, the following reaction occurs[1]:

$$3NaHCO_{3(s)} + H_3C_6H_5O_{7(s)} + H_2O_{(l)} \rightarrow C_6H_5O_7Na_{3(aq)} + 4H_2O_{(l)} + 3CO_{2(g)}$$

Design an experiment to study the effect of temperature on the reaction rate of Alka Seltzer™ in water. Check with the instructor regarding how many Alka Seltzer™ tablets are available for each team.

PROCEDURE

NOTE: Consult safety information and obtain teacher approval before beginninng the experiment. Design an experiment to determine the relationship between temperature and the rate of the reaction.

SAFETY INFORMATION

1. Do not stopper or cover the container in which the reaction is taking place.
2. Keep the temperature of the water between 5°C and 85°C.
3. After reviewing your procedure, the instructor will discuss any safety precautions that are specific to your experiment.

QUESTIONS FOR FURTHER THOUGHT

1. What are some other variables that could affect the rate of this reaction?
2. What factors might make one type of effervescent tablet more effective than another?

REFERENCE

1. Faulkner, S. P. *The Science Teacher.* 1993, 60(1), 26–29.

15 The Effect of Temperature on the Reaction Rate of Alka Seltzer™ in Water

TEACHER'S NOTES

Alka Seltzer™ tablets provide a relatively safe and inexpensive way for students to study many aspects of reactions. In this lab, students will investigate the effect of temperature on the rate of the reaction.

OPTIONS

1. This experiment can be done in addition to, or in place of, experiment 14, The effect of temperature on the rate of a clock reaction. If the chemicals required to perform experiment 14 are not available, this reaction is an alternative that provides reliable data.
2. If you do not want to spend a lot of time on this experiment, have each team pick a different temperature to study and then have the class pool their results.
3. Students could also investigate the effect of changing surface area on the rate of this reaction.
4. Instead of only investigating the role temperature plays in this reaction, you could let the students compete to see who can get the tablet to react in the shortest amount of time. In this case, students can vary temperature, volume of water, surface area of the tablet, and so on.

NOTES

1. Students must first decide on the volume of water to use in each trial. If they use extremely small volumes, the reaction may not go to completion. Students can learn this through trial and error, or you may suggest that they use a minimum of 100 mL of water.
2. Students may have trouble deciding when the reaction has finished. Some teams wait until the tablet has stopped fizzing, while other teams will continue timing the reaction until the tablet has completely dissolved. The important thing is that the teams be consistent in determining what constitutes the end of this reaction.
3. Effervescent tablets other than Alka Seltzer™ can be used. Tablets can be purchased through many chemical supply companies.
4. You may want to provide each team with one tablet to use for a trial run.

TIME

The amount of time needed for students to design and execute their own experiment can vary depending on the level and creativity of the class, their familiarity with designing their own investigations, and the complexity of the procedure they develop.

Average Time for This Experiment

Time needed to design the experiment: 15 minutes
Time needed to run the experiment: 30 minutes

TEAMS

Students may conduct the experiment individually or work in teams of 2 depending on the number of tablets available.

MATERIALS (PER TEAM)

Require the use of goggles, gloves, and aprons/lab coats. The following supplies should be readily available for the students:

- Alka Seltzer™ tablets (3–4)
- 250-mL or 400-mL beakers
- Thermometers
- Hot plate or other heat source
- Ice
- Stopwatch
- Tongs
- 100-mL graduated cylinder
- Mortar and pestle

SAMPLE PROCEDURES

1. Students may add one tablet to a given amount of water that has been heated or cooled to various temperatures and time the reaction.
2. Students might grind the tablets in a mortar and pestle and add a certain mass of tablet to a given amount of water that has been heated or cooled to various temperatures and time the reaction. This ensures that each trial uses the same mass of tablet.
3. Students might determine the mass of the tablets ahead of time to determine if all tablets have the same mass. If the mass varies from tablet to tablet, they may convert the data to sec/g of tablet.

SAFETY PRECAUTIONS

NOTE: Also refer to the safety information given with the experiment. Discuss with the students any safety precautions that are specific to the procedure they develop. Material Safety Data Sheets for each substance used should be reviewed and made available to the students.

COMMON MISCONCEPTIONS, PROCEDURAL AND CALCULATION ERRORS

1. Students might assume that all tablets are identical.
2. Students might use different thermometers for each trial.
3. Students might add ice to a given amount of water to cool it below room temperature and then not remeasure the volume of the water before adding the tablet. This will result in an increased volume of water due to the melted ice.

LAB REPORT

There are several alternatives to writing a formal lab report given in the introduction. If students are writing their own lab report it should include:

I. Title
II. Purpose
III. Materials

IV. Procedure
V. Data
VI. Calculations and/or graph
VII. Conclusion (including a discussion of sources of error)
VIII. Answers to Questions (optional)

ANSWERS TO THE QUESTIONS

1. Other factors that might affect the rate of this reaction include the surface area of the tablet, the volume of water, or the purity of the water.
2. The quantity and quality of active ingredients could make one type of tablet more effective than another. The rate of the reaction shouldn't be a major factor in its effectiveness because the person does not take the medicine until the reaction is over.

FOR FURTHER READING

Faulkner, S.P. *The Science Teacher.* 1993, 60(1), 26–29.

15 The Effect of Temperature on the Reaction Rate of Alka Seltzer™ in Water

SAMPLE LAB REPORT

PURPOSE

To determine the effect of temperature on the reaction rate of Alka Seltzer™ in water.

MATERIALS

Safety equipment: goggles, apron and gloves
100-mL graduated cylinder
Thermometer
Hot plate
Ice
Stopwatch
Alka Seltzer™ tablets (4)
Tongs
250-mL beaker

PROCEDURE

1. Add 100.0 mL of water to the 250-mL beaker and record the temperature of the water.
2. Add an Alka Seltzer™ tablet to the water and time the reaction.
3. Rinse the solution down the sink.
4. Pour another 100.0 mL of water into the 250-mL beaker and heat the water to approximately 20°C above room temperature.
5. Remove the beaker from the heat.
6. Record the temperature of the water.
7. Add an Alka Seltzer™ tablet to the water and time the reaction.

8. Rinse the solution down the sink.
9. Repeat steps 4–8 with water that is heated to approximately 40°C above room temperature.
10. Add another 100 mL of water to the beaker and three to four ice cubes. Stir until the temperature of the water is about 8°C.
11. Remove the remaining ice cubes with the tongs.
12. Pour the water into the graduated cylinder to obtain 100 mL of cooled water. Pour the remaining water down the sink.
12. Pour the 100 mL of cooled water into the beaker and record its temperature.
13. Add an Alka Seltzer™ tablet to the cooled water and time the reaction.

DATA

TEMPERATURE (°C)	TIME (sec)
10.2	123.64
17.1	57.92
35.3	37.22
60.1	23.86

CALCULATIONS AND/OR GRAPH

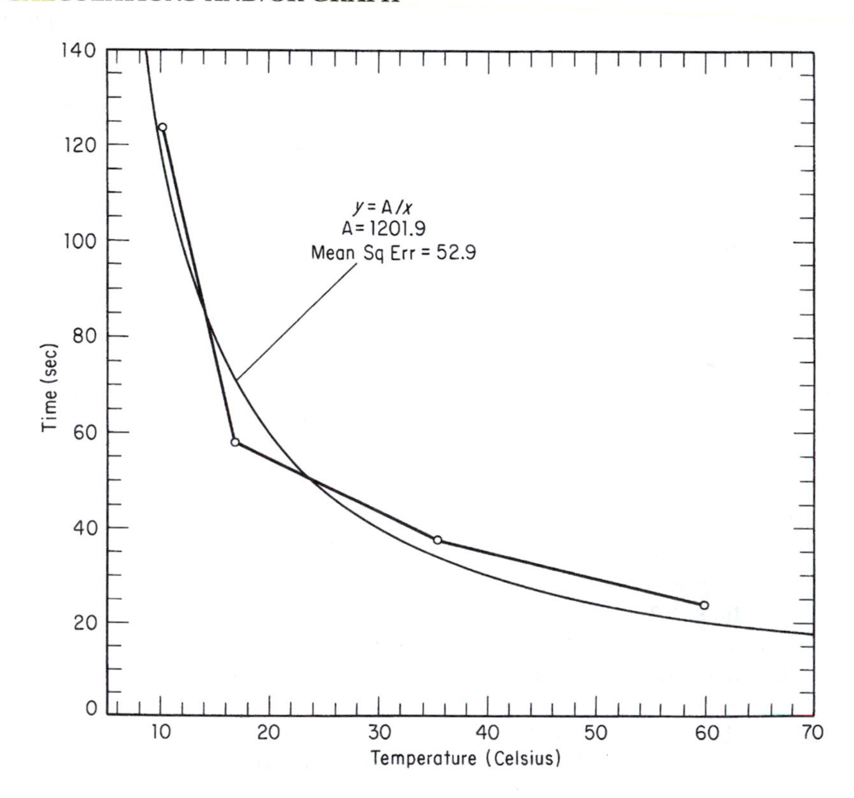

CONCLUSION

The data show that as the temperature of the water increased, the rate of the reaction also increased. The graph shows both the data and the best-fit curve. This demonstrates that the two variables may be inversely proportional. The equation of the best-fit curve gives the relationship between the variables as: time (sec) = 1200/temp. (°C).

SOURCES OF ERROR

The following is a compilation of sources of error that students may suggest:

1. The accuracy of the thermometer is unknown.
2. There isn't any way to be sure if each tablet is identical in its composition.
3. The temperature of the water may change during the course of the reaction.
4. It was difficult to clearly identify when the reaction was finished.

16 Calculating the Heat of Solution

THE EXPERIMENT

PURPOSE

To determine the heat of solution of a known solute.

BACKGROUND

The heat of solution for a given solute is the quantity of heat that is absorbed (endothermic) or released (exothermic) as a solute dissolves in a given amount of water per mole of the solute.

The unit for heat of solution is kcal/mol or kJ/mol in a given quantity of water (although kcal/g or kJ/g is also acceptable). The symbol used for heat of solution is ΔH_{soln}. If the dissolving process is endothermic, the heat of solution is given as a positive value ($+\Delta H_{soln}$). If the dissolving process is exothermic, the heat of solution is given as a negative value ($-\Delta H_{soln}$).

PROCEDURE

NOTE: Consult safety information and obtain teacher approval before beginning the experiment. You will be given a sample of a known solute such as anhydrous calcium chloride or sodium thiosulfate pentahydrate. The instructor will specify the mass of solute available. Design an experiment to determine the quantity of heat that is gained or lost by the solute as it dissolves in a given sample of water and calculate its heat of solution.

SAFETY INFORMATION

1. Sodium thiosulfate pentahydrate is a white deliquescent crystal. In case of contact with skin, wash with large quantities of water. If contact with the inside of the mouth occurs, wash the mouth and consult a physician. For contact with the eyes, wash eyes for 15 minutes and see a physician.
2. Calcium chloride can cause severe irritation and burns when it comes in contact with the skin.
3. The instructor will alert the class to safety information for other solutes.

4. After reviewing your procedure, the instructor will discuss any safety precautions that are specific to your experiment.

QUESTIONS FOR FURTHER THOUGHT

1. If the dissolving process of a given solute is endothermic, as the solute dissolves will the temperature of the water increase or decrease?
2. The heat of solution of sodium nitrate is 5.03 kcal/mol solute in 200 g of water. Calculate the quantity of heat that would be absorbed if 50.0 g of sodium nitrate dissolves in 200 g of water. What would be the change in the temperature of the water?
3. A type of instant cold compress sold on the market contains ammonium nitrate and water. When the inner bag of ammonium nitrate is broken, the crystals dissolve in water, making the bag feel cold. What does this indicate about the sign of the heat of solution of ammonium nitrate (i.e., is it a positive value or a negative value)?

16 Calculating the Heat of Solution

TEACHER'S NOTES

In this lab, students are asked to calculate the heat of solution of a known solute. If students have not performed one of the experiments using the concepts and equations involved in calorimetry, you may want to spend time discussing these aspects of the lab before the students begin. If you would like more detailed background information regarding calorimetry experiments, refer to the teacher's notes for experiment 7, The heat of fusion of ice.

OPTIONS

1. Each team can be given either the anhydrous calcium chloride or the sodium thiosulfate pentahydrate.
2. Each team can be given both a sample of the anhydrous calcium chloride and the sodium thiosulfate pentahydrate to show the difference in the sign of the ΔH_{soln}.

NOTES

1. Anhydrous calcium chloride or sodium thiosulfate pentahydrate work well as the solutes in this experiment. Calcium chloride has a negative heat of solution (exothermic), and sodium thiosulfate pentahydrate has a positive heat of solution (endothermic). Giving some teams the anhydrous calcium chloride and some teams the sodium thiosulfate can work well to demonstrate the $-\Delta H$ and $+\Delta H$, respectively.
2. Providing each team with 6–7 g of the solute supplies a large enough sample to perform a few trials and still vary the mass of solute used per trial.
3. Students are given some flexibility in the introduction regarding the unit to use for heat of solution. You may want to specify a particular unit.
4. Students are not told how much water to use in the experiment. You may specify the volume to use (100 mL works well) or leave the decision to the students. If the class is making a group decision regarding the quantity of water, ask the students to consider the pros and cons of using large or small quantities of water. Some factors that students may suggest include:

   A. The larger the quantity of water used, the smaller the temperature change (given the same amount of solute). Students choosing to use quantities like 500 mL of water may find the change in temperature too small to be measured with any accuracy. If this occurs, they should realize that lowering the volume of the water will result in a greater ΔT, given the same mass of solute, and make the adjustment in their procedure.

   B. If the quantity of water chosen by the students is too small, they may not be able to get all of the solute to dissolve. Both calcium chloride and sodium thiosulfate pentahydrate are very soluble in water, but volumes <25.0 mL may not be adequate (depending on the mass of solute they decide to use).

5. Some students may decide to use 1 g of solute in a trial run to check its solubility in a given amount of water.

6. Another factor that you may want students to think about is whether or not to stir the solution as the crystals are dissolving. If students use small quantities of solute (1-g samples) and allow the crystals to sit and dissolve slowly, the heat being absorbed or released from the crystals may be slowly transferred to or from the air, producing a very small ΔT. However, if students stir the solution to increase the rate of solution and cause a more rapid change in heat and a more measurable ΔT, the stirring itself may cause an increase in the exchange of heat between the water and the air. Some students may decide to crush the crystals before adding them to the water to increase the rate of dissolving without stirring, thereby attempting to minimize the exchange of heat between the water and the air.

TIME

The amount of time needed for students to design and execute their own experiment can vary depending on the level and creativity of the class, their familiarity with designing their own investigations, and the complexity of the procedure they develop.

Average Time for This Experiment

Time needed to design the experiment: 10–15 minutes (if students are familiar with calorimetry)

Time needed to run the experiment: 30–40 minutes

TEAMS

Teams of two works well.

MATERIALS (PER TEAM)

Require the use of goggles, gloves, and aprons/lab coats. The following supplies should be readily available for the students:

- Solute (such as anhydrous calcium chloride or sodium thiosulfate pentahydrate): 6–7 g
- Polystyrene cups (8 or 16 oz)
- Beakers
- Thermometers
- Balance
- Weighing paper or weighing dishes
- 100-mL graduated cylinder
- Stirring rod
- Distilled water (optional)

SAFETY PRECAUTIONS

NOTE: Also refer to the safety information given with the experiment. Discuss with the students any safety precautions that are specific to the procedure they develop. Material Safety Data Sheets for each substance used should be reviewed and made available to the students.

COMMON MISCONCEPTIONS, PROCEDURAL AND CALCULATION ERRORS

1. Students may forget that they should calculate the change in heat of the water (ΔH_{water}) and then assume that the ΔH_{solute} is numerically equal but opposite in sign.
2. Students may calculate the change in heat (ΔH_{solute}) but not adjust the value to be in terms of 1.0 mol.
3. Students may peform the experiment in a beaker rather than in an insulated container.
4. Small sample sizes or large quantities of water can produce a very small change in the temperature of the water.
5. Students may use the mass of the crystals when calculating ΔH_{water}.

LAB REPORT

NOTE: There are several alternatives to writing a formal lab report given in the introduction. If students are writing a formal lab report it should include:

I. Title
II. Purpose
III. Materials
IV. Procedure
V. Data
VI. Calculations
VII. Conclusion (including a discussion of sources of error)
VIII. Answers to Questions (optional)

ANSWERS TO THE QUESTIONS

1. If the dissolving process of a given solute is endothermic, the temperature of the water will decrease as the solute dissolves because the solute will absorb heat from the water.
2. 50.0 g $NaNO_3$ = 0.588 mol $NaNO_3$

$$0.588 \text{ moles} \times \frac{5.03 \text{ kcal}}{1 \text{ mol}} = 2.96 \text{ kcal of heat would be absorbed by the crystals.}$$

If 2.96 kcal of heat is absorbed by the $NaNO_3$, then 2.96 kcal of heat would be lost by the water.

Using the ΔH equation: $\Delta H = (\text{mass})(\Delta T)(\text{specific heat})$

$$2.96 \text{ kcal} = (200 \text{ g})(\Delta T)(0.001 \text{ kcal/g·°C})$$

$$\Delta T = 14.8°\text{C}$$

Therefore, the temperature of the water would decrease 14.8°C.

3. Because the dissolving process of ammonium nitrate in water decreases the temperature of the water, it can be concluded that the dissolving process is endothermic and therefore the heat of solution would have a positive value.

16 Calculating the Heat of Solution

SAMPLE LAB REPORT

PURPOSE

To determine the heat of solution of anhydrous calcium chloride.

MATERIALS

Safety equipment: goggles, gloves, and apron
Calcium choride crystals (6 g)
Polystyrene cup
100-mL graduated cylinder
Thermometer
Weighing dish
Stirring rod
Balance

PROCEDURE

1. Obtain approximately 6 g of anhydrous calcium chloride.
2. Measure 100 mL of tap water and pour it into a polystyrene cup.
3. Record the temperature of the tap water in the cup.
4. Measure approximately 2.00 g of anhydrous calcium chloride on the balance.
5. Add the crystals to the water and stir.
6. Monitor the temperature of the water in the cup. If the temperature of the water increases, record the maximum temperature. If the temperature of the water decreases, record the minimum temperature.
7. Repeat the experiment using new samples of anhydrous calcium chloride.

DATA

	TRIAL		
	1	2	3
Mass of anhydrous calcium chloride (g)	2.01	1.51	2.23
Initial temperature of water (°C)	19.7	19.2	20.2
Final temperature of water (°C)	22.4	21.4	23.5

CALCULATIONS

Trial 1

Change in temperature of the water: $22.4 - 19.7 = 2.7°C$

Change in heat of water: $\Delta H = m \times \Delta T \times c$

$= 100.0 \text{ g} \times 2.7°C \times 4.18 \text{ J/g·°C}$

$= 1100 \text{ J} = 1.1 \text{ kJ}$

Heat lost by the crystals $= -1.1$ kJ

Moles of anhydrous calcium chloride used: $2.01 \text{ g} \times 1 \text{ mol}/110.98 \text{ g} = 0.0181 \text{ mol}$

Heat of solution of anhydrous calcium chloride = ΔH/mol = −1.1 kJ/0.0181 mol = −61 kJ/mol

Trial 2

Change in temperature of the water: 21.4 − 19.2 = 2.2°C
Change in heat of water: $\Delta H = m \times \Delta T \times c$
$= 100.0 \text{ g} \times 2.2°\text{C} \times 4.18 \text{ J/g·°C}$
$= 920 \text{ J} = 0.92 \text{ kJ}$
Heat lost by the crystals = −0.92 kJ
Moles of anhydrous calcium chloride used: 1.51 g × 1 mol/110.98 g = 0.0136 mol
Heat of solution of anhydrous calcium chloride = ΔH/mol = −0.92 kJ/0.0136 mol = −67 kJ/mol

Trial 3

Change in temperature of the water: 23.5°C − 20.2°C = 3.3°C
Change in heat of water: $\Delta H = m \times \Delta T \times c$
$= 100.0 \text{ g} \times 3.3°\text{C} \times 4.18 \text{ J/g·°C}$
$= 1400 \text{ J} = 1.4 \text{ kJ}$
Heat lost by the crystals = −1.4 kJ
Moles of anhydrous calcium chloride used: 2.23 g × 1 mol/110.98 g = 0.0201 mol
Heat of solution of anhydrous calcium chloride = ΔH/mol = −1.4 kJ/0.0201 mol = −70 kJ/mol
Average value: (−61 kJ/mol) + (−67 kJ/mol) + (−70 kJ/mol) /3 = −66 kJ/mol

CONCLUSION

Based on our data, the heat of solution of anhydrous calcium chloride is −66 kJ/mol in 100 g of water.

SOURCES OF ERROR

The following is a compilation of sources of error that students may suggest:

1. The crystals may not be pure.
2. Impurities in the tap water may interfere with the results. Using distilled water instead of tap water may reduce this error.
3. Some of the heat from the crystals may have been lost to the air. Covers for the cups may have reduced this error.
4. Loss of heat to the air through the cup. If we had used a second cup as an insulator, we might have been able to reduce this error.
5. The initial temperature of the water was not room temperature, so the water may have gained or lost heat to the air. We could have waited for the water to reach room temperature before we started.
6. We stirred the solution to increase the rate of dissolving, but this may have increased the rate of heat exchange between the water and the air.

17 Determining the Molarity of a Saturated Solution

THE EXPERIMENT

PURPOSE

To determine the molarity of a saturated solution at room temperature.

BACKGROUND

In this experiment, you will be given a sample of a known solute (your instructor will identify the solute to be used). Design an experiment to determine the molarity of a saturated solution. Molarity (M) is a unit used to indicate the concentration of a solution. The equation for molarity is:

Molarity = moles of solute/liter of solution

PROCEDURE

NOTE: Consult safety information and obtain teacher approval before beginning the experiment. Design a procedure to prepare a saturated solution and calculate the molarity of the saturated solution.

SAFETY INFORMATION

After reviewing your procedure, the instructor will discuss any safety precautions that are specific to your experiment.

QUESTIONS FOR FURTHER THOUGHT

1. Calculate the molarity of a solution that contains 17.54 g of KCl in 2.00 L of solution.
2. What mass of $C_6H_{12}O_6$ would be required to prepare 2.00 L of a 1.50 M solution?
3. What is the relationship between temperature and the molarity of a saturated NaCl solution?

17 Determining the Molarity of a Saturated Solution

TEACHER'S NOTES

In this experiment, students are asked to calculate the molarity of a saturated solution. In experiment 18, students are asked to calculate the molality of a saturated solution.

This lab can be used when students first learn how to convert to moles, or when studying solubility, units of concentration, or solubility products.

OPTIONS

1. The solute that the student(s) will be using is not specified in the lab, so you can use what is available. Recommended solutes are NaCl, KCl, sucrose (common table sugar), or dextrose (glucose). The entire class can use the same solute, or they can be divided among the various teams. If more than one solute will be used, the students can be asked to predict the relative solubilities of the solids before beginning the lab.

2. The class can be divided in half, with one half performing this experiment and the other half of the class performing experiment 18, Calculating the molality of a saturated solution. If this option is preferred, it can be beneficial for the students to compare their procedures and calculations to emphasize the difference between the units of molarity and molality and how they are calculated. (The same solutes can be used in both labs.)

NOTES

1. According to the *Merck Index,*[1] the recommended solutes have approximately the following solublities at room temperature:

SOLUTE	g/100 mL OF SATURATED SOLUTION[1]	MOLARITY
NaCl	31.7	5.42
KCl	31.2	4.18
$C_{12}H_{22}O_{11}$ (sucrose)	90.9	2.66
$C_6H_{12}O_6$ (glucose)	59	3.28

2. At the conclusion of the lab, it is interesting for students to be given the information in the chart above so they can compare the relative solubilities of the solutes in terms of grams and moles and then discuss which unit is more valuable to scientists and why.

3. Students who decide to let their undissolved solute dry overnight will need some time another day to determine their final mass measurement.

4. Emphasize to the students that there has to be some evidence that the solution was saturated.

5. You may want to remind the students that the more water they use, the greater the quantity of solute required to form a saturated solution. Students using large volumes of water can deplete the stock of the various solutes very quickly. To avoid this pitfall, you may want to limit the volume of water that can be used to 100 mL.

6. Most high school texts have a solubility curve that shows the solubility of various solutes over a range of temperatures. If the students are using one of the solutes shown on the graph, they can approximate the mass of solute that will be required to form a saturated solution.

TIME

The amount of time needed for students to design and execute their own experiment can vary depending on the level and creativity of the class, their familiarity with designing their own investigations, and the complexity of the procedure they develop.

Average Time for This Experiment

Time needed to design the experiment: 30 minutes
Time needed to run the experiment: 60–90 minutes (can be divided into two lab periods)

TEAMS

Teams of two to three work well.

MATERIALS (PER TEAM)

Require the use of goggles, gloves, and aprons/lab coats. The following supplies should be readily available for the students:

- Solute (~50 g)
- Stirring rod
- Beaker
- Rubber policeman
- Graduated cylinder
- Weighing paper or weighing dishes
- Balance
- Filter paper
- Evaporating dish and watch glass
- Dropper pipet
- Funnel
- Heat source (hot plate or Bunsen burner)

SAMPLE PROCEDURES

1. THE FOLLOWING PROCEDURE IS NOT RECOMMENDED: Students may want to add scoops of solute to a significant volume of water (100–500 mL) until excess solute appears on the bottom of the beaker. Then they might attempt to filter the solution and boil the filtrate to dryness to isolate the dissolved salt. Due to the large quantities of water and salt often involved in students' procedures, there are potential hazards asso-

ciated with the boiling process. You may want to suggest method #4 below or recommend that they only use a small volume of water (perhaps 10 mL). This will reduce the risks and save time as well.

2. Some students may mass small amounts of solute (1-g samples, for example) and add each sample of the solute to a given amount of water until excess solute appears on the bottom of the beaker. Then they filter the solution, measure the volume of the filtered solution, and determine the mass of the excess solute (after drying overnight). If students leave the filtered solution overnight, it should be covered to reduce evaporation.
3. Some students might mass small amounts of solute (1-g samples, for example) and add each sample of the solute to a given amount of water until excess solute appears on the bottom of the beaker. Then they add water to the mixture, perhaps 1 mL at a time, until all of the solute has dissolved and measure the volume of the final solution.
4. In all of the previous procedures, students used the entire solution for their data and calculations. Some students may realize that they can simply calculate the molarity of a sample of the saturated solution. These students will saturated a sample of water, pour off a given volume of the solution (perhaps 10 mL), and evaporate the sample in an evaporating dish (either overnight or with a heat source) to obtain the mass of solute present in the 10-mL sample of saturated solution.
5. The lab does not specify that the solution must be saturated at room temperature. Some students may add solute to the water until the solution is saturated with excess solute remaining on the bottom of the beaker and then heat the solution slowly, recording the temperature at which all of the solute dissolves. Then, they calculate the molarity of the saturated solute solution at x°C. If you do not want this to be an option, specify that the saturated solution must be at room temperature.
6. Some students may mass samples of solute and add them to a given amount of water until saturation occurs. Then they repeat the procedure, adding a little less solute in the last step. If some solute still settles out, they repeat the procedure using even less solute in the last step. The students repeat the process until only a few grains of solute settle out in the beaker. This method can be very time consuming.
7. Students that filter the excess solute may decide to let it dry overnight, while others may try to scrape it into an evaporating dish and heat it to drive off the water.

SAFETY PRECAUTIONS

Discuss with the students any safety precautions that are specific to procedure they develop. Material Safety Data Sheets for each substance used should be reviewed and made available to the students.

The main safety precautions for this experiment involve the teams that decide to boil the solution to obtain the mass of solute that has dissolved. If students attempt to boil 100 mL or more of saturated solution to dryness, the large quantities of salt found in the solutions can produce potential hazards during the boiling process (see the recommendations for procedure #1 in the previous section).

COMMON MISCONCEPTIONS, PROCEDURAL AND CALCULATION ERRORS

1. Students might think that the volume measurement used in calculating molarity is the volume of water added to the mixture rather the volume of the resulting solution. Students should use the volume of the saturated *solution* in their calculation.
2. Students may not take into account the mass of the excess solute in the beaker.
3. Students might use the *mass* of solute that dissolved when they calculate the molarity of the solution rather than converting to *moles* of solute that dissolved.

LAB REPORT

There are several alternatives to writing a formal lab report given in the introduction. If students are writing a formal lab report, it should include the following:

I. Title
II. Purpose
III. Materials
IV. Procedure
V. Data
VI. Calculations
VII. Conclusion (including a discussion of sources of error)
VIII. Answers to Questions (optional)

ANSWERS TO THE QUESTIONS

Note: These questions are similar to those asked in experiment 18 in the event that half of the class is doing this experiment while the rest are working on experiment 18.

1. (17.54 g KCl) (1 mol KCl/74.56 g) = 0.2352 mol KCl
   0.2352 mol KCl/2.00 L = 0.118 M KCl solution
2. (1.50 mol glucose/L solution) (2.00 L solution) = 3.00 mol glucose
   (3.00 mol glucose)(180 g glucose/1 mol) = 540 g glucose
3. The solubility of NaCl increases as the temperature of the solution increases. Therefore, the molarity of a saturated solution of NaCl also increases with increasing temperature.

REFERENCES

1. *Merck Index,* 9th ed., Windholz, M. (ed.), Merck and Co., Inc., Rahway, NJ, 1976.

Acknowledgment Adapted with permission from the *Journal of Chemical Education,* 71 (9), 1994, 792–793, copyright 1994, Division of Chemical Education, Inc.

17 Determining the Molarity of a Saturated Solution

SAMPLE LAB REPORT

PURPOSE

To determine the molarity of a saturated NaCl solution.

MATERIALS

Safety equipment: goggles, gloves, and apron
NaCl
600-mL beaker
Stirring rod
Filter paper
Weighing dish
100-mL graduated cylinder
Balance
Funnel
Rubber policeman

PROCEDURE

1. Pour 200 mL of water into a 600-mL beaker.
2. Add salt, 5 g at a time, until no more will dissolve.
3. Record the total mass of NaCl added to the water.
4. Record the mass of the filter paper.
5. Filter the solution through filter paper using a rubber policeman to get all of the excess salt onto the filter paper.
6. Record the volume of the filtered solution.
7. Allow the filter paper and excess salt to dry overnight.
8. Record the mass of the filter paper and excess salt when dry.

DATA

Initial volume of water: 200.0 mL
Mass of NaCl added to water: 75.00 g
Mass of filter paper: 1.02 g
Volume of filtered solution: 223.0 mL = 0.223 L
Mass of filter paper and excess salt: 10.77 g

CALCULATIONS

Mass of excess NaCl in sample: 10.77 g − 1.02 g = 9.75 g NaCl
Mass of NaCl that dissolved: 75.00 g − 9.75 g = 65.25 g NaCl
Moles of NaCl dissolved: 65.25 g × 1 mol NaCl/58.45 g = 1.116 mol NaCl
Molarity of saturated solution = 1.116 mol NaCl/0.223 L = 5.01 M

CONCLUSION

Based on our data, the molarity of a saturated solution of NaCl at room temperature is 5.01 M.

SOURCES OF ERROR

The following is a compilation of sources of error that students may suggest:

1. Some water from the solution was absorbed by the filter paper, which changed the final volume of the solution.
2. We used tap water, and the impurities could have remained behind with the dried salt.
3. The NaCl may have contained impurities.
4. The filter paper and excess salt may still have been wet.
5. Table salt is slightly deliquescent and may have absorbed moisture from the air, changing its mass.

18 Determining the Molality of a Saturated Solution

THE EXPERIMENT

PURPOSE

To determine the molality of a saturated solution.

BACKGROUND

In this experiment, you will be given a sample of a known solute (your instructor will identify the solute to be used). Design an experiment to determine the molality of a saturated solution. Molality (m) is a unit used to indicate the concentration of a solution. The equation for molality is:

Molality = moles of solute/kilogram of solvent

PROCEDURE

NOTE: Consult safety information and obtain teacher approval before beginning the experiment. Design an experiment to prepare a saturated solution of a given solute and calculate the molality of the saturated solution.

SAFETY INFORMATION

After reviewing your procedure, the instructor will discuss any safety precautions that are specific to your experiment.

QUESTIONS FOR FURTHER THOUGHT

1. Calculate the molality of a solution that contains 20.00 g of KCl dissolved in 300 g of water.
2. What mass of glucose ($C_6H_{12}O_6$) would be required to prepare a 3.00 m solution in 2.50 kg of water?
3. What is the relationship between temperature and the molality of a saturated NaCl solution?

18 Determining the Molality of a Saturated Solution

TEACHER'S NOTES

In the experiment, students are asked to calculate the molality of a saturated solution. In experiment 17, students are asked to calculate the molarity of a saturated solution.

This lab can be used when students are first learning how to convert grams to moles, or when they are studying solutions, units of concentration, or solubility products.

OPTIONS

1. The solute that the students will be using is not specified in the lab, so you can use what is available. Recommended solutes are NaCl, KCl, sucrose (common table sugar), or dextrose (glucose). The entire class can use the same solute, or they can be divided among the various teams. If more than one solute will be used, the students can be asked to predict the relative solubilities of the solids before beginning the lab.

2. The class can be divided in half, with one half performing this experiment and the other half of the class performing experiment 17, Calculating the molarity of a saturated solution. If this option is preferred, it can be beneficial for the students to compare their procedures and calculations to emphasize the difference between the units of molarity and molality and how they are calculated. (The same solutes can be used in both labs.)

NOTES

1. The molality of the various solutes at room temperature is given below. These values were calculated according to the formula and values given in the *Handbook of Chemistry and Physics.*[1] These are approximate values for a saturated solution at room temperature.

SOLUTE	MOLALITY
NaCl	6.1
KCl	4.2
$C_{12}H_{22}O_{11}$ (sucrose)	15.3
$C_6H_{12}O_6$ (glucose)	8.3

2. Students who decide to let their undissolved solute dry overnight will need some time another day to determine their final mass measurement.

3. Emphasize to the students that there has to be some evidence that the solution was saturated.

4. You may want to remind the students that the more water they use, the more solute that will be required to form a saturated solution. Students using large volumes of water can deplete the stock of solutes very quickly. You may want to limit the volume of water that can be used.

TIME

The amount of time needed for students to design and execute their own experiment can vary depending on the level and creativity of the class, their familiarity with designing their own investigations, and the complexity of the procedure they develop.

Average Time for This Experiment

Time needed to design the experiment: 30 minutes
Time needed to run the experiment: 60–90 minutes (can be divided into two lab periods)

TEAMS

Teams of two to three work well.

MATERIALS (PER TEAM)

Instructors should require the use of goggles, gloves, and aprons/lab coats. The following supplies should be readily available for the students:

- Solute
- Stirring rod
- Beakers
- Graduated cylinder
- Weighing paper or weighing dishes
- Balance
- Filter paper
- Evaporating dish and watchglass
- Dropper pipet
- Funnel
- Distilled water
- Heat source (hot plate or Bunsen burner)

SAMPLE PROCEDURES

1. Students may add a small amount of solute (1 g samples, for example) to a given amount of water until excess salt remains on the bottom of the beaker, after which they may:

   A. Filter the excess solute and allow it to dry overnight to determine the mass of solute that did not dissolve,
   B. Filter the excess solute and boil the remaining saturated solution to dryness. (This should only be done with small samples [5–10 mL] of saturated solution. See safety precautions below.)
   C. Repeat the entire process, except adding less solute in the last step to closer approximate saturation, repeating as necessary, or
   D. Add water to the solution 1 mL at a time until all of the solute dissolves.

2. Some students may saturate a sample of water with solute and then pour about 10 mL of the saturated solution into an evaporating dish. They obtain the mass of the sample and boil off the water to determine the mass of solute in the sample.

3. If students have studied colligative properties of solutions, they may saturate a sample of water and then pour a sample of the saturated solution into a beaker and determine its freezing point or boiling point. Then stu-

dents use the formula $\Delta T_{b\ or\ f} = (K_{b\ or\ f})(m)$ to determine the molality of the saturated solution.

SAFETY PRECAUTIONS

Discuss with the students any safety precautions that are specific to the procedure they develop. Material Safety Data Sheets for each substance used should be reviewed and made available to the students.

The main safety precaution involves boiling a large sample of saturated solution to dryness. Boiling a large volume of saturated solution involves a large mass of solute, which can create potential hazards and should be avoided. Encourage students designing this type of procedure to use small volumes of water (5–10 mL).

COMMON MISCONCEPTIONS, PROCEDURAL AND CALCULATION ERRORS

1. Students may forget that 1.00 mL of water = 1.00 g.
2. When initially designing their experiment, students may not realize that they need to account for the mass of excess solute remaining in the saturated solution.
3. When performing their calculations, students may use the mass of solute added to the solution rather than converting to moles of solute.

LAB REPORT

There are several alternatives to writing a formal lab report given in the introduction. If students are writing their own lab report it should include:

I. Title
II. Purpose
III. Materials
IV. Procedure
V. Data
VI. Calculations
VII. Conclusion (including a discussion of sources of error)
VIII. Answers to Questions (optional)

ANSWERS TO THE QUESTIONS

NOTE: The questions asked in this lab are similar to those asked in experiment 17 in the event that half of the class is doing this experiment while the rest are working on experiment 17.

1. 20.00 g KCl × 1 mol/74.56 g = 0.2682 mol KCl
   300 g water = 0.300 kg water
   Molality of the solution = 0.2682 mol KCl /0.300 kg water = 0.894 m solution
2. 2.50 kg water × 3.00 mol glucose /kg of water = 7.50 mol glucose
   7.50 mol glucose × 180 g/1 mol = 1.35×10^3 g glucose
3. As the temperature of a NaCl solution increases, the solubility of NaCl also increases. Therefore, at higher temperatures, more NaCl can be dissolved, which will increase the molality of the solution.

REFERENCE

1. Weast, R.C., ed. *Handbook of Chemistry and Physics,* 57th ed., Chemical Rubber Company, Cleveland, OH, 1976, pp. D-218–261.

Acknowledgment Adapted with permission from the *Journal of Chemical Education,* 71 (9), 1994, 792–793; copyright 1994, Division of Chemical Education, Inc.

18 Determining the Molality of a Saturated Solution

SAMPLE LAB REPORT

PURPOSE

To determine the molality of a saturated NaCl solution.

MATERIALS

Safety equipment: goggles, gloves, and apron
NaCl
Hot plate
250-mL beaker
Weighing dish
Evaporating dish
Watchglass to fit evaporating dish
Balance
Stirring rod

PROCEDURE

1. Pour 100 mL of water into the 250-mL beaker.
2. Add 40 g of NaCl to the beaker. Stir to dissolve as much salt as possible. If all of the NaCl dissolves add another 10-g sample.
3. Continue until excess NaCl remains on the bottom of the beaker.
4. Let the excess NaCl settle to the bottom of the beaker.
5. Determine the mass of the evaporating dish and watchglass.
6. Pour 10 mL of the solution into the evaporating dish.
7. Determine the mass of the evaporating dish, watchglass, and saturated solution.
8. Cover the evaporating dish with the watchglass.
9. Heat the evaporating dish on the hot plate until all of the water has evaporated.
10. Let the evaporating dish cool and obtain the mass of the dish, watchglass, and salt.

DATA

Mass of the empty evaporating dish and watch glass: 36.41 g
Mass of the evaporating dish, watchglass, and saturated solution: 46.15 g
Mass of evaporating dish, watchglass, and NaCl: 38.88 g

CALCULATIONS

Mass of the solution: 46.15 g − 36.41 g = 9.74 g
Mass of water in solution: 46.15 g − 38.88 g = 7.27 g = 0.00727 kg
Mass of NaCl: 9.74 g − 7.27 g = 2.47 g

$$\text{Moles of NaCl: } 2.47 \text{ g NaCl} \times \frac{1 \text{ mole}}{58.45 \text{ g}} = 0.0423 \text{ mol NaCl}$$

$$\text{Molality of saturated solution} = \frac{0.0423 \text{ mol NaCl}}{0.0072 \text{ kg water}} = 5.88 \text{ m}$$

CONCLUSION

The molality of a saturated solution of NaCl at room temperature is 5.88 m.

SOURCES OF ERROR

The following is a compilation of sources of error that students may suggest:

1. The salt we used may not be completely pure.
2. We used tap water, so the impurities in the water may have remained in the evaporating dish after boiling.
3. Some of the salt may have been lost during the boiling process.
4. The solution wasn't saturated.

19 Factors Affecting the Rate of Solution

THE EXPERIMENT

PURPOSE

To determine what factors affect the rate of solution.

BACKGROUND

Rate of solution is the amount of time needed for a given amount of solute to dissolve in a given amount of solvent. For this experiment you will be given approximately 5 g of sodium thiosulfate pentahydrate ($Na_2S_2O_3 \cdot 5H_2O$) and will study the effect of several variables on the rate of solution.

PROCEDURE

NOTE: Consult safety information and obtain teacher approval before beginning the experiment.

1. List three or four variables you believe affect the rate of solution.
2. Write a hypothesis stating how each of these three or four variables will affect the rate at which sodium thiosulfate crystals dissolve in water. (Will the rate of solution increase or decrease?)
3. Design a series of experiments to study the effect of each of the variables on the rate of solution (i.e., rate of dissolving).

SAFETY INFORMATION

1. Sodium thiosulfate pentahydrate is a white deliquescent crystal.
2. In the event of external contact, wash the affected areas with large quantities of water.
3. For contact with eyes, wash continuously for 15 minutes and call a physician.
4. After reviewing your procedure, the instructor will discuss any safety precautions that are specific to your experiment.

QUESTIONS FOR FURTHER THOUGHT

1. Describe the process of NaCl dissolving in water.
2. Cite some specific examples where the rate of dissolving for a particular solute would be an important factor.

19 Factors Affecting the Rate of Solution

TEACHER'S NOTES

In this experiment, students are asked to choose three or four variables that they believe will affect the rate of solution of sodium thiosulfate pentahydrate in water. Students typically choose the following variables to study: (1) temperature of the solvent, (2) surface area of the crystal, (3) volume of solvent (water) used, and (4) stirring

OPTIONS

1. You can specify to the students the three or four variables you want them to study.
2. The entire class can discuss what variables they think will affect the rate of solution, and then each team can study one of the factors. At the conclusion of the lab, each team can report their results to the class.
3. Different teams can use different substances (such as NaCl, sugar, or Alka Seltzer™ tablets) to see if they achieve the same general results.
4. You could give each team 1 g of sodium thiosulfate pentahydrate and have a competition to determine which team can get the crystals to dissolve in the shortest amount of time. Students should incorporate several of the variables mentioned above.

NOTES

1. Students should include a control in their experiment.
2. If the temperature of the water is going to be studied as a variable, students should be asked to keep the temperature of the water between 5°C and 80°C. Sodium thiosulfate pentahydrate crystals should not be heated above 100°C.
3. Students may have some difficulty identifying the point at which all of the crystals have dissolved.
4. The 5 g sample of sodium thiosulfate crystals allows the students to perform five trials (one control and four experimental) using 1 g per trial. If teams design experiments that test more than four variables, you may want to have more crystals available.
5. Students may need some guidance deciding what volume of water to use because they will not know the solubility of sodium thiosulfate pentahydrate in water. The crystals are fairly soluble, allowing students to use 50 mL of water or more per trial. You could give the students an extra 1 g sample to determine the volume of water to use in each trial.
6. When studying the effect of temperature on the rate of solution, students should heat (or cool) the water before adding the crystals.
7. Because the crystals are deliquescent, you may want to distribute 1-g samples as needed or encourage the students to keep the containers tightly covered.

TIME

The amount of time needed for students to design and execute their own experiment can vary depending on the level and creativity of the class, their familiarity with designing their own investigations, and the complexity of the procedure they develop.

Average Time For This Experiment

Time need to design the experiment: 20 minutes
Time needed to run the experiment: 30–40 minutes

TEAMS

Teams of two to three work well.

MATERIALS (PER TEAM)

Require the use of goggles, gloves, and aprons/lab coats. The following supplies should be readily available for the students:

- 250-mL beakers
- Stirring rod
- Stopwatches (or clock)
- Mortar and pestle
- Heat source
- Ice
- Balance
- Weighing paper
- Thermometer
- Scoop or spoon

SAMPLE PROCEDURE

Many students will develop a procedure in which they take a given amount of crystal and dissolve it in a given amount of water, while timing the dissolving process. Then they repeat the procedure, only this time they:

1. Heat or cool the water first,
2. Crush the crystals in a mortar and pestle before adding them to the water,
3. Add the crystals and stir, or
4. Change the volume of water.

SAFETY PRECAUTIONS

NOTE: Also refer to the safety information given with the experiment. Discuss with the students any safety precautions that are specific to the procedure they develop. Material Safety Data Sheets for each substance used should be reviewed and made available to the students.

Avoid contact between sodium thiosulfate pentahydrate and strong oxidizers and acids. Do not heat above 100°C, as decomposition can release toxic fumes.

COMMON MISCONCEPTIONS, PROCEDURAL AND CALCULATION ERRORS

1. Students might not include a control in their experiment.
2. Students might change more than one variable in a given trial.

LAB REPORT

There are several alternatives to writing a formal lab report given in the introduction. If students are writing a formal lab report it should include:

I. Title
II. Purpose
III. Hypothesis
IV. Procedure
V. Data
VI. Conclusion (including a discussion of sources of error)
VII. Answers to Questions (optional)

ANSWERS TO THE QUESTIONS

1. When NaCl is added to water, the Na^+ ions of the salt are attracted to the negative (oxygen) end of the polar water molecules, and the Cl^- ions of the salt are attracted to the positive (hydrogen) end of the water molecules. The water molecules surround the ions of the salt, isolating them from the other charged ions. The ions of the salt are now said to be hydrated. The hydrated ions move away from the crystal and become dispersed throughout the solution.
2. Possible answers include dissolving medicine in water or in the stomach where an increased rate of dissolving may mean that the medicine will work sooner, cleaning products and the rate at which they dissolve stains, a toxic spill and the rate of dissolving into the ground water.

19 Factors Affecting the Rate of Solution

SAMPLE LAB REPORT

PURPOSE

To determine what factors affect the rate of solution.

HYPOTHESIS

Our hypothesis is that heating the water will increase the rate of solution, stirring will increase the rate of solution, crushing the crystals will increase the rate of solution, and increasing the volume of solvent will increase the rate of solution.

MATERIALS

Safety equipment: goggles, gloves, and apron
5 g of sodium thiosulfate pentahydrate
Stirring rod
Mortar and pestle
250-mL beaker
Hot plate
Stopwatch
Thermometer

PROCEDURE

1. Obtain 1 g of sodium thiosulfate pentahydrate.
2. Put 100 mL of water into a 250-mL beaker.

3. Add the crystals to the water and time how long it takes for them to completely dissolve.
4. Obtain another 1-g sample of crystals.
5. Crush the crystals in a mortar and pestle.
6. Add the crushed crystals to another 100 mL of water and time how long it takes for them to dissolve completely.
7. Obtain 1 g of crystals.
8. Add the crystals to another 100 mL of water and stir continuously. Time how long it takes for them to dissolve completely.
9. Obtain 1 g of crystals.
10. Add the crystals to 200 mL of water and time how long it takes for them to dissolve completely.
11. Heat 100 mL of water to 60°C.
12. Obtain 1 g of crystals.
13. When the water reaches 60°C, add the crystals and time how long it takes for them to dissolve completely.

DATA

	CONTROL	CRUSHED CRYSTALS	STIRRING	MORE SOLVENT	HEATED SOLVENT
Time to dissolve (min. : sec.)	5:45	1:33	1:20	3:57	2:51

CONCLUSION

Our hypothesis was correct. Each of the following variables increased the rate of solution: (1) crushing the crystals, (2) stirring the water, (3) adding more water, (4) heating the water. (Some students may make a bar graph of the data or rank the variables in terms of their relative effect. Students may suggest why each variable affected the rate of solution.)

SOURCES OF ERROR

The following is a compilation of sources of error that students may suggest:

1. Some of the crystals stayed in the mortar and pestle, so that trial may not have had 1.00 g of solute.
2. Sometimes it was difficult to determine if all of the crystals had dissolved.
3. The crystals are deliquescent, so they may have been absorbing moisture from the air, and that could change the mass measurement.
4. We may not have used exactly the same amount of water each time.
5. The temperature of the tap water may have changed from one trial to the other and we didn't measure it each time.

20 Calculating the Quantity of Heat Released from The Heat Solution™ Handwarmer

THE EXPERIMENT

PURPOSE

To calculate the quantity of heat released by The Heat Solution™ handwarmer.

BACKGROUND

The Heat Solution™ handwarmer contains a supersaturated solution of sodium acetate trihydrate and a stainless steel disc. When the metal disc is flexed, the excess sodium acetate that is in solution will begin to recrystallize. The recrystallization of sodium acetate trihydrate is an exothermic process and will therefore release heat. The pouch can be reused by placing it in boiling water for 10 minues and then allowing it to cool slowly.

Design an experiment to calculate the quantity of heat released by the handwarmer.

PROCEDURE

NOTE: Consult safety information and obtain teacher approval before beginning experiment.

When you are ready to begin the experiment, *gently* flex the disc in the pouch, the sodium acetate will begin to crystallize (the bag will begin to fill with white crystals), and heat will immediately begin being released. Be careful not to bend the disc to the point that it could break. Return the pouch to the instructor when the experiment is completed.

SAFETY INFORMATION

1. Squeezing the pouch before or after crystallization has occurred can cause the bag to rupture. If any solution (or crystals) come in contact with skin, rinse immediately with running water for 15 minutes and notify the instructor. If any solution (or crystals) leak onto the lab table, notify the instructor. NOTE: If the pouch does begin to leak, it may produce an odor similar to vinegar.

2. If sodium acetate comes in contact with the eyes, rinse in an eyewash for 15 minutes and call a physician.
3. Overhandling The Heat Solution™ pouch can cause the recrystallization process to begin prematurely. Set the pouch aside until you are ready to begin the experiment.
4. The reaction produces enough heat that the potential for burns exists. Use caution.
5. Be careful not to bend the disc to the point that it could break.
6. After reviewing your procedure, the instructor will discuss any safety precautions that are specific to your experiment.

QUESTIONS FOR FURTHER THOUGHT

1. How is a supersaturated solution prepared?
2. Is the dissolving of sodium acetate trihydrate an endothermic or exothermic process?

20 Calculating the Quantity of Heat Released from The Heat Solution™ Handwarmer

TEACHER'S NOTES

In this experiment students are asked to determine the quantity of heat released by the handwarmer. This lab can be completed while the students are studying the properties of solutions or heat and calorimetry. If the students have not performed one of the experiments that uses calorimetry, you may want to review the concepts and equations involved and refer to the teacher's notes for experiment 7, The heat of fusion of ice.

OPTIONS

This lab can be performed in conjunction with the other experiments based on calorimetry (when studying heat and energy) or when the students are studying solutions and the concept of heat of solution. If a lot of time has passed between when the students completed the other calorimetry experiments and when they are working on this lab, you may need to spend more time reviewing the concepts and equations involved.

If a limited number of handwarmers is available, this lab can be done in conjunction with another experiment. For example, the class could be divided in half with some teams working on the handwarmer experiment and other teams working on experiment 8, The heat of crystallization of wax. During the next lab period they can switch experiments.

NOTES

1. The Heat of Solution™ handwarmers are reusable. The solution in the pouch can be returned to the state of supersaturation by placing it in boiling water until all of the crystals have redissolved (approximately 10 minutes). Remove the pouch from the water and allow it to cool to room temperature. Because of the time required to return the pouch to supersaturation, students will probably only be able to complete one trial, unless a large number of handwarmers is available. The pouch can be reused approximately 30–40 times.[1]
2. The Heat Solution™ handwarmers are sold by Flinn Scientific, Inc., P.O. Box 219, 131 Flinn Street, Batavia, IL 60510; (708) 879-6900. Each handwarmer costs approximately $6.50.
3. If you have enough handwarmers available, you may want to demonstrate to the students how to activate the pouch.
4. Emphasize to the students that they should not overhandle the pouch before the experiment or after it has been activated.
5. The equation for the reversible reaction is:[2]

$$NaC_2H_3O_{2(l)} + \text{(at least 3) } H_2O_{(l)} \leftrightarrow NaC_2H_3O_2{\cdot}3H_2O_{(s)} + 19.7 \text{ kJ}$$

6. The vinyl pouch will remain hot for about 30 minutes.
7. The maximum temperature reached by the pouch is 58°C or 136°F.[3]

8. If the students say that they smell vinegar, one of the pouches has probably ruptured.
9. The use of other handwarmers is not recommended.

TIME

NOTE: The amount of time needed for students to design and execute their own experiment can vary depending on the level and creativity of the class, their familiarity with designing their own investigations, and the complexity of the procedure they develop. The times below are for students familiar with investigations involving calorimetry.

Average Time for this Experiment

Time needed to design the experiment: 15 minutes
Time needed to run the experiment: 20–30 minutes

TEAMS

The number of students assigned per team will depend on the number of handwarmers available. Teams of two work well.

MATERIALS (PER TEAM)

Require the use of goggles, gloves, and aprons/lab coats. The following supplies should be readily available for the students:

- The Heat Solution™ handwarmers
- Styrofoam cups (16-oz. cups or larger work well. The handwarmers cannot fit in smaller cups and still be submerged)
- 600-mL beakers (some teams will not think about using an insulated container and will want a beaker)
- Thermometers (alcohol thermometers work fine)
- Balances (some groups may want to determine the mass of the handwarmer and calculate the quantity of heat lost by the handwarmer per gram)
- 100-mL graduated cylinders

SAFETY PRECAUTIONS

NOTE: Also refer to the safety information given with the experiment. Discuss with the students any safety precautions that are specific to the procedure they develop. Material Safety Data Sheets for each substance used should be reviewed and made available to the students. If the pouch leaks, wash the lab area (or skin) with running water.

COMMON MISCONCEPTIONS, PROCEDURAL AND CALCULATION ERRORS

1. Students may forget that the density of water is 1.00 g/mL.
2. Students may forget the formula for calculating the change in heat: $\Delta H = m \times \Delta T \times c$, where m mass (of water), ΔT = change in temperature (of the water), and c = specific heat of water (1.00 cal/g·°C or 4.18 J/g·°C)

3. Students may mistakenly use the mass of the handwarmer as m in the ΔH equation rather than the mass of the water in the cup.

4. Students may not ask for a Styrofoam cup to minimize heat loss.

5. Students may activate the handwarmer in the air rather than waiting until it is submerged. Waiting to activate the handwarmer until after it is submerged minimizes the quantity of heat lost to the air.

LAB REPORT

NOTE: There are several alternatives to writing a formal lab report given in the introduction. If students are writing a formal lab report, it should include the following:

I. Title
II. Purpose
III. Materials
IV. Procedure
V. Data
VI. Calculations
VII. Conclusion (including a discussion of sources of error)
VIII. Answers to Questions (optional)

ANSWERS TO THE QUESTIONS

1. A supersaturated solution is prepared by creating a saturated solution at a high temperature and then allowing the solution to cool slowly. If the excess solute remains dissolved, the solution will be supersaturated.
2. The recrystallization of sodium acetate was exothermic (the pouch released heat while the solute was crystallizing), and therefore dissolving (the reverse process) must be endothermic.

REFERENCES

1. *Chem Fax*™. Flinn Scientific, Inc., Publication no. 1933, Batavia, IL, 1990.
2. Ibid.
3. Ibid.

FOR FUTHER READING

Daley, H.O. Jr. *Journal of Chemical Education.* 1994, 71, 791–792.

Shakhashiri, B.Z. *Chemical Demonstration: A Handbook for Teachers of Chemistry,* vol. 1. University of Wisconsin Press, Madison, WI, 1983, pp. 27–30.

20 Calculating the Quantity of Heat Released from The Heat Solution™ Handwarmer

SAMPLE LAB REPORT

PURPOSE

To calculate the quantity of heat released by a handwarmer.

MATERIALS

Heat Solution handwarmer
Styrofoam cup
Thermometer
100-mL graduated cylinder

PROCEDURE

1. Measure 400 mL of tap water in the graduated cylinder.
2. Pour the water into the Styrofoam cup.
3. Record the temperature of the water.
4. Gently place the handwarmer in the water, making sure it is completely submerged.
5. Reach into the cup and gently flex the disc, being careful not to spill any water from the cup. Shake excess water from your hand back into the cup.
6. Using the thermometer, try to stir the water as much as possible to distribute the heat evenly throughout the cup.
7. Record the maximum temperature reached by the water.
8. Return the handwarmer to the instructor.

DATA

Volume of water in the cup: 400.0 mL
Mass of water in the cup: 400.0 g
Initial temperature of the water in the cup: 22.0°C
Final temperature of the water in the cup: 30.0°C

CALCULATIONS

Change in temperature of the water (ΔT) = 30.0°C − 22.0°C = 8.0°C

Change in heat of the water: ΔH_{water} = (mass of water)(ΔT)(specific heat of water)
= (400.0 g)(8.0°C)(1.00 cal/g·°C)
= 3200 calories
= 3.2 10^3 calories

Change in heat of water = change in heat of handwarmer (assumed)

$\Delta H_{water} = \Delta H_{handwarmer}$
$= 3.2 \times 10^3$ calories

CONCLUSION

The quantity of heat lost by the handwarmer is 3.2×10^3 calories.

SOURCES OF ERROR

The following is a compilation of the sources of error that students may suggest:

1. Some heat was lost to the environment between when the pouch was activated and when it was placed in the water. (For teams who activate the pouch outside of the water.)
2. Some of the heat released into the water was transferred to the air at the surface.
3. Water may have been lost from the cup when activating the pouch under the water.
4. We might not have waited long enough to get the final temperature of the water.
5. The heat released by the handwarmer was not evenly distributed throughout the water. When the thermometer was held close to the pouch, there was a greater change in temperature.
6. The water in the cup was not stirred enough to distribute the heat evenly.
7. The handwarmer was not completely submerged in the water and some heat was lost to the air.
8. The volume of water was not measured accurately.
9. The accuracy of the thermometer is not known.

21 The Effect of Concentration on the Boiling Point of a Sugar Solution

THE EXPERIMENT

PURPOSE

To study the effect of molal concentration on the boiling point of a sugar solution.

BACKGROUND

The addition of a solute to a solvent produces several changes in the properties of that solvent. Dissolving a solute in a solvent will change the vapor pressure, freezing point, and boiling point of the solvent. These are colligative properties because they depend only on the number of, and not the identify of, the particles in the solution.

In this experiment, you will determine the relationship between the molal concentration of a sugar (sucrose) solution and its boiling point. The formula for sucrose (common table sugar) is $C_{12}H_{22}O_{11}$.

PROCEDURE

NOTE: Consult safety information and obtain teacher approval before beginning the experiment. Design an experiment to determining the relationship between the molal concentration of a sucrose solution and its boiling point.

SAFETY INFORMATION

1. When using a hot plate as the heat source, be careful not to spill any sugar onto the surface of the hot plate, as it will immediately burn. Remove the beaker from the heat source before adding sugar.
2. After reviewing your procedure, the instructor will discuss any safety precautions that are specific to your experiment.

QUESTIONS FOR FURTHER THOUGHT

1. What effect does concentration have on the vapor pressure and freezing point of a solution?

2. Why is the antifreeze used in cars also "antiboil"? What is the solute in antifreeze? Why is it important to periodically check the concentration of antifreeze in the radiator of an automobile?

21 The Effect of Concentration on the Boiling Point of a Sugar Solution

TEACHER'S NOTES

In this experiment, students investigate the effect of a solution's molal concentration on its boiling point. This experiment can be performed when students are studying properties of solutions, units of concentration, or colligative properties.

You can use this lab to confirm the relationship between the variables after studying the concept in class or use it as an introduction to colligative properties.

The lab is intentionally written with the simple purpose of investigating the relationship between molal concentration and boiling point. However, with some classes you may want specify that the students are to determine a *mathematical* relationship between these two variables. Students could be asked to determine if molal concentration and boiling point are directly proportional or to calculate the value of the boiling point constant: 0.52°C/m.

OPTIONS

1. Students could be asked to specifically determine a mathematical relationship between the variables or to calculate the boiling point constant.
2. Students could repeat the experiment using NaCl instead of sugar if you want them to understand the nature of colligative properties (i.e., that adding an ionic solute that dissociates in water will have twice [in this case] or three times the effect of adding a nonelectrolyte like sucrose, per mole of solute).
3. Students could also investigate the relationship between molal concentration and the freezing point of a sucrose solution.

NOTES

1. Students should realize that all of the sugar that they add to their predetermined volume of water must dissolve. You could provide the students with some extra sugar to test for solubility.
2. After the first trial, when students realize that the boiling point of the solution increases, some teams may ask if they can just add more sugar to the existing "boiling" solution rather than restarting with room temperature water each time. This will save a lot of time in the experiment, but will not account for the water lost during boiling and therefore a change in concentration. The decision is left to your discretion.
3. To save time, each student on the team can run one of the trials, but should use the same thermometer to measure the final boiling point of the solution.
4. The larger the sample of water used in each trial, the longer it will take for the solution to reach the boiling point. You may want to discourage students from using large samples. A maximum of 250 mL of water is recommended.

5. Beakers of sugar can be left at each station ahead of time or given to each team after their procedure has been approved.

TIME

The amount of time needed for students to design and execute their own experiment can vary depending on the level and creativity of the class, their familiarity with designing their own investigations, and the complexity of the procedure they develop.

Average Time for This Experiment

Time needed to design the experiment: 15–20 minutes
Time needed to run the experiment: 30–50 minutes

The run time depends on whether the students continue to add more sugar to the same boiling solution, whether they divide solution samples among team members, and the volume of water being used. (Teams using 400 mL of water in each trial will not be able to finish in this amount of time.)

TEAMS

Teams of three work well.

MATERIALS (PER TEAM)

Require the use of goggles, gloves, and aprons/lab coats. The following supplies should be readily available for the students:

- Sucrose (common table sugar works fine)
- Hot plate or other heat source
- 250 mL–600 mL beakers
- Weighing dishes or weighing paper
- Balance
- Stirring rod
- Thermometer
- 100-mL graduated cylinder

SAMPLE PROCEDURES

Many students will design an experiment in which they add three or four different masses of sugar to a given amount of water and measure the boiling point of the solution. As previously mentioned, some teams may continue to add sugar to the same boiling solution, while others will start each trial with room-temperature water.

Some teams might vary the concentration of the solution by holding the quantity of sugar constant and varying the volume of water used in each trial.

SAFETY PRECAUTIONS

NOTE: Also refer to the safety information given with the experiment. Discuss with the students any safety precautions that are specific to the procedure they develop. Material Safety Data Sheets for each substance used should be reviewed and made available to the students.

COMMON MISCONCEPTIONS, PROCEDURAL AND CALCULATION ERRORS

1. Students might not record the boiling point of the pure water and therefore not be able to determine the ΔT of the boiling point.
2. Students might assume the boiling point of the pure water to be 100°C and not check the calibration of the thermometer.
3. Students might forget that 1.00 mL of water = 1.00 g = 1.0×10^{-3} kg. This may give them some trouble when they try to convert the volume of water to kilograms.
4. Some students might think that they can't calculate molality because they did not measure the mass of the water and the formula for molality involves kilograms of solvent.

LAB REPORT

There are several alternatives to writing a formal lab report given in the introduction. If students are writing their own lab report it should include:

I. Title
II. Purpose
III. Materials
IV. Procedure
V. Data
VI. Calculations and/or Graphs
VII. Conclusion (including a discussion of sources of error)
VIII. Answers to Questions (optional)

ANSWERS TO THE QUESTIONS

1. Adding a solute to a solvent will decrease the vapor pressure of the solvent as well as the freezing point.
2. Antifreeze is also antiboil because it lowers the freezing point and raises the boiling point of the water in the radiator. The most common solute in antifreeze is ethylene glycol. It is important to periodically check the concentration of the antifreeze because if it becomes too dilute it will not protect the engine to the same extent, and the water in the radiator could either freeze or boil, causing damage to the engine.

21 The Effect of Concentration on the Boiling Point of a Sugar Solution

SAMPLE LAB REPORT

PURPOSE

To study the effect of molal concentration on the boiling point of a sugar solution.

MATERIALS

- Safety equipment: goggles, gloves, and apron
- Sucrose
- 400-mL beakers (2)
- Hot plate
- Balance
- Weighing dish
- 100-mL graduated cylinder
- Thermometer
- Stirring rod

PROCEDURE

1. Measure 200 mL of water into a 400-mL beaker.
2. Heat the water until it begins to boil and measure the boiling point.
3. Measure 200 mL of water into another 400-mL beaker and add 10 g of sugar.
4. Heat the solution until it begins to boil and measure the boiling point.
5. Repeat steps 3 and 4 with 30 g of sugar and 50 g of sugar.

DATA

SOLUTION	MASS OF SUGAR ADDED (g)	BOILING POINT OF SOLUTION (°C)
A	0.00	99.6
B	10.00	101.5
C	30.00	102.7
D	50.00	103.1

CALCULATIONS

SOLUTION	MASS OF SUGAR (g)	MOLES OF SUGAR (MW = 342 g)	MOLALITY (moles sugar/ 0.200 kg)	CHANGE IN BOILING POINT (°C)
A	0.00	0.00	0.00	—
B	10.00	0.0292	0.146	1.9
C	30.00	0.0877	0.439	3.1
D	50.00	0.146	0.730	3.5

CONCLUSION

There is a correlation between the concentration of a sugar solution and its boiling point. Adding sugar to water raises the boiling point. As the concentration of the solution increased, the boiling point also increased.

SOURCES OF ERROR

The following is a compilation of sources of error that students may suggest:

1. The sugar may not be completely pure.
2. We didn't use distilled water.

3. The thermometer was resting on the bottom of the beaker, and therefore the temperature readings may not be accurate.
4. We spilled some sugar while adding it to the beaker.
5. As the water boiled, the concentration of the solution continued to change due to the evaporation of water.

22 Determining the Melting Point of Lauric Acid

THE EXPERIMENT

PURPOSE

To determine the melting point of lauric acid.

BACKGROUND

The melting point of a substance is the temperature at which the substance changes from a solid to liquid. During melting the kinetic energy of the substance remains constant while the energy absorbed is used to increase the potential energy of the substance.

PROCEDURE

NOTE: Consult safety information and obtain teacher approval before beginning the experiment. Design an experiment to determine the melting point of lauric acid.

SAFETY INFORMATION

1. Melt the acid by placing the test tube in a warm water bath. The bath should be approximately 60°C. Do not directly heat the test tube.
2. Remove the stopper from the test tube before heating.
3. Do not heat the acid above 60°C.
4. The thermometer can be placed in the sample of lauric acid.
5. To perform several trials, the substance can be "refrozen" by placing the test tube in a beaker of tap water.
6. If the thermometer becomes frozen in the solid lauric acid, place the test tube in a warm water bath and wait for the acid to melt. Do not attempt to remove the thermometer from the solid.
7. Do not pour the liquid acid in the sink. It is not water soluble.
8. After completing the experiment, return the test tube to the instructor.
9. Wipe the thermometer thoroughly after completing the lab.

10. After reviewing your procedure, the instructor will discuss any safety precautions that are specific to your experiment.

QUESTIONS FOR FURTHER THOUGHT

1. What is the freezing point of lauric acid?
2. Based on the melting point of lauric acid, would you predict that the attractive forces between the particles are relatively weak or strong compared to other solids? Why?
3. Would the mass of the sample affect the value obtained for the melting point?

22 Determining the Melting Point of Lauric Acid

TEACHER'S NOTES

In this experiment, students are provided with a sample of lauric acid and asked to determine its melting point. This lab can be used when students are studying phase changes, bonding, and the effect that the strength of the intermolecular forces has on melting point, or the concepts of kinetic and potential energy.

OPTIONS

1. The use of lauric acid is recommended due to its low melting point, but you may use other compounds if you are familiar with their risks.
2. You can ask some students to determine the melting point of the substance while others determine the freezing point. Ideally, students will arrive at the same value.
3. You can ask all of the students to determine both the melting and freezing point of the acid.
4. If the chemistry lab has temperature probes available that will not be affected by the lauric acid, students can use a probe with a CBL or computer interface to monitor and graph their data. The direct-connect temperature probe sold by Vernier[1] can be used with lauric acid.
5. If students will be graphing the data, you can ask them to indicate on the graph where the potential energy and kinetic energy of the acid were changing.

NOTES

1. Lauric acid (or dodecanoic acid) has the formula $CH_3(CH_2)_{10}COOH$. It has the appearance of colorless needles or white flakes and the odor of candle wax. Lauric acid is insoluble in water; it is soluble in ether and benzene. The melting point of lauric acid is 43–45°C.
2. Place approximately 8 g of lauric acid in individual test tubes and stopper. The samples can be reused for several years.
3. Students obtain better data if the warm water bath is approximately 60°C. If the water becomes very hot (80–90°C), the temperature may not plateau.
4. If students design an experiment in which they try to obtain the melting point of the lauric acid when they initially melt the substance, they usually won't get very good results. Ideally, students should resolidify the solid in a beaker of tap water, allowing the thermometer to become frozen inside, and then repeat the melting process.
5. To clean the thermometer at the conclusion of the lab, students will probably have to use a paper towel that has been dipped in warm water.

TIME

The amount of time needed for students to design and execute their own experiment can vary depending on the level and creativity of the class, their familiarity with designing their own investigations, and the complexity of the procedure they develop.

Average Time for This Experiment

Time needed to design the experiment: 20 minutes
Time needed to run the experiment: 40 minutes

TEAMS

Teams of two work well.

MATERIALS (PER TEAM)

Require the use of goggles, gloves, and aprons/lab coats. The following supplies should be readily available for the students:

- Test tube of lauric acid (or another compound)
- Heat source
- 250- or 400-mL beakers (2)
- Thermometer (or temperature probe)
- Test tube holder
- Stopwatch or clock

SAMPLE PROCEDURES

Students will probably design an experiment in which they place the test tube of lauric acid in a warm water bath and monitor the temperature as the substance is warmed and melts. Students might continue by placing the test tube in a bath of tap water to resolidify the acid to determine its freezing point. The students might then repeat the melting process and record the temperature changes again.

The main differences among student procedures are:

1. How often they take the temperature of the acid,
2. How many trials they perform,
3. Whether or not they stir the sample as it is being heated or cooled, and
4. Whether or not they record temperatures during heating and cooling to compare the melting and freezing points.

SAFETY PRECAUTIONS

This lab should be performed in a fume hood.

NOTE: Also refer to the safety information given with the experiment. Discuss with the students any safety precautions that are specific to the procedure they develop. Material Safety Data Sheets for each substance used should be reviewed and made available to the students.

COMMON MISCONCEPTIONS, PROCEDURAL AND CALCULATION ERRORS

Students may not understand that during a phase change the temperature of the substance remains constant because the potential energy (rather than the kinetic energy) of the material is changing.

LAB REPORT

There are several alternatives to writing a formal lab report given in the introduction. If students are writing their own lab report it should include:

I. Title
II. Purpose
III. Materials
IV. Procedure
V. Data
VI. Graph (optional)
VII. Conclusion (including a discussion of sources of error).
VIII. Answers to Questions

ANSWERS TO THE QUESTIONS

1. The freezing point of lauric acid is equal to the melting point.
2. The melting point of the acid was found to be 45°C. This is a low melting point, compared to other solids, and therefore would indicate that the attractive forces between the molecules of lauric acid are relatively weak compared to other solids.
3. The mass of the sample does not affect the melting point of a substance. Melting point is an intensive property.

REFERENCES

1. Vernier Software, 8565 S.W. Beaverton-Hillsdale Hwy., Portland, OR 97225-2429; telephone (503) 297-5317.

22 Determining the Melting Point of Lauric Acid

SAMPLE LAB REPORT

PURPOSE

To determine the melting point of lauric acid.

MATERIALS

Safety equipment: goggles, gloves, and apron
Test tube with lauric acid
250-mL beakers (2)
Hot plate
Test tube holder
Stopwatch
Thermometer

PROCEDURE

1. Fill the two 250-mL beakers approximately ¾ full with tap water.
2. Begin heating one of the beakers on the hot plate. Place a thermometer

in the water. Heat the water to 60°C and then turn off the heat source. Remove the beaker from the hot plate.

3. Remove the stopper from the test tube of lauric acid.
4. Place the thermometer in the test tube.
5. Place the test tube in the warm water bath. Begin stirring when the acid starts to melt.
6. As soon as the acid is completely melted, remove the sample from the warm water.
7. Place the test tube in the beaker of tap water.
8. Record the temperature of the lauric acid every 10 seconds until the lauric acid has reached a temperature of 35.0°C. NOTE: Allow the thermometer to become frozen in the acid. Do not attempt to remove the thermometer from the frozen lauric acid.
9. Check the temperature of the warm water bath. Reheat the water until it has returned to 60°C. Place the test tube in the warm water bath.
10. Record the temperature of the lauric acid every 10 seconds as it melts.
11. Clean the thermometer with a paper towel and restopper the test tube.
12. Return the test tube to the instructor.

DATA

COOLING DATA

TIME (sec)	TEMP (°C)	TIME (sec)	TEMP (°C)
0	52.6	300	42.1
10	52.4	310	41.7
20	51.8	320	41.4
30	50.9	330	41.1
40	50.3	340	40.5
50	49.6	350	40.1
60	49.5	360	39.4
70	49.1	370	39.3
80	48.3	380	38.6
90	47.9	390	38.1
100	46.3	400	37.1
110	45.6	410	36.7
120	44.4	420	36.5
130	44.4	430	36.4
140	44.1	440	36.2
150	44.1	450	36.1
160	44.1	460	35.7
170	44.1	470	35.3

HEATING DATA

TIME (sec)	TEMP (°C)	TIME (sec)	TEMP (°C)
0	24.4	300	44.2
10	24.6	310	44.3
20	24.7	320	44.3
30	25.3	330	44.3
40	27.2	340	44.5
50	28.6	350	44.5
60	29.8	360	44.5
70	30.1	370	44.5
80	32.5	380	44.8
90	33.3	390	44.9
100	34.3	400	45.2
110	35.1	410	45.6
120	36.3	420	45.9
130	36.6	430	46.3
140	37.4	440	46.5
150	38.4	450	46.9
160	39.1	460	47.1
170	39.4	470	47.8

COOLING DATA

TIME (sec)	TEMP (°C)	TIME (sec)	TEMP (°C)
180	44.1	480	35.1
190	43.9	490	34.7
200	43.9	500	34.5
210	43.8	510	34.2
220	43.7	520	33.7
230	43.7	530	32.9
240	43.6	540	32.1
250	43.4	550	31.8
260	43.1	560	31.4
270	43.1	570	31.1
280	42.7	580	30.7
290	42.5	590	30.5

HEATING DATA

TIME (sec)	TEMP (°C)	TIME (sec)	TEMP (°C)
180	39.9	480	48.2
190	40.6	490	48.6
200	40.8	500	48.9
210	41.4	510	49.2
220	41.8	520	49.7
230	42.2	530	49.9
240	42.7	540	50.3
250	42.8	550	50.8
260	43.1	560	51.6
270	43.4	570	52.8
280	43.8	580	53.4
290	44.2	590	54.1

GRAPH

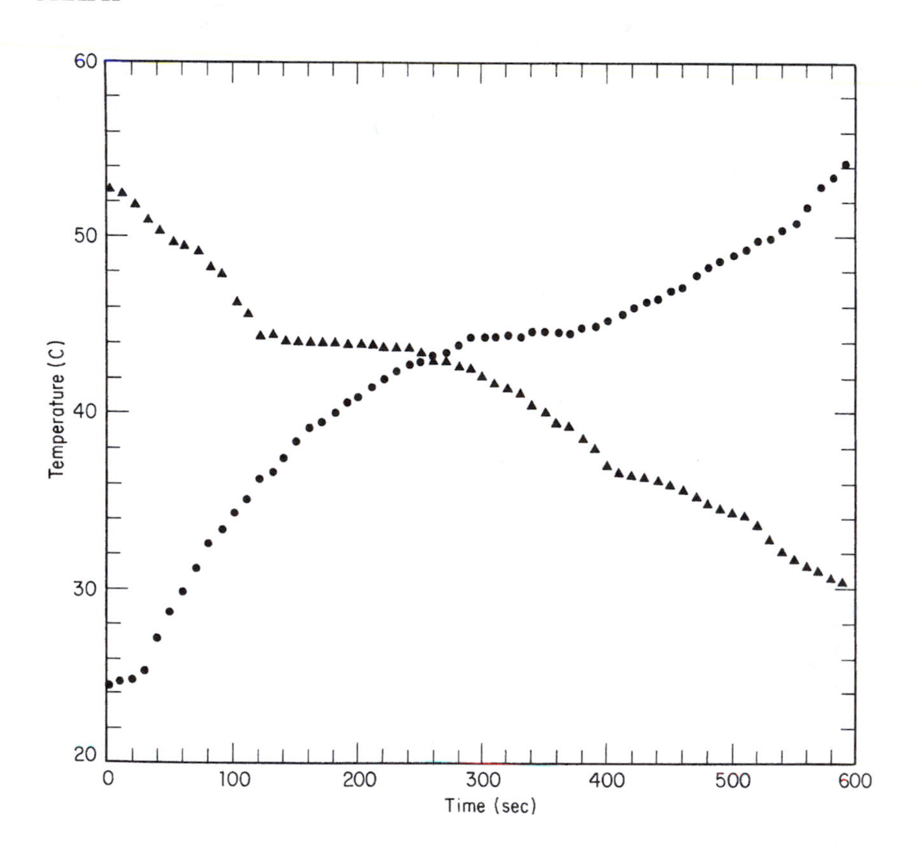

CONCLUSION

On the cooling curve, the temperature of the acid seems to plateau at 44.1°C, while on the warming curve it seemed to remain steady at 44.3°C. Ideally, the plateaus should occur at the same temperature on each graph, but the differences could be due to sources of error. From the data, our conclusion is that the melting point of lauric acid is approximately 44°C.

SOURCES OF ERROR

The following is a compilation of sources of error that students may suggest:

1. The sample of lauric acid may not be pure.
2. Water may have splashed into the sample.
3. The thermometer may not be calibrated correctly.

23 The Effect of Temperature on the Reaction Time of Magnesium and Vinegar

THE EXPERIMENT

PURPOSE

To determine the effect of temperature on the reaction time of magnesium and vinegar.

BACKGROUND

The length of time over which a reaction takes place is affected by many factors, including the strength of the bonds being broken or formed, the activation energy of the reaction, the presence or absence of a catalyst, the surface area of a solid reactant, the pressure of a gaseous reactant, the concentration of aqueous reactant, and the temperature.

The purpose of this experiment is to determine the relationship between temperature and the time needed for the following reaction to be completed:

$$Mg_{(s)} + 2HC_2H_3O_{2(aq)} \rightarrow Mg(C_2H_3O_2)_{2(aq)} + H_{2(g)}$$

For the experiment, you will be given approximately 5 cm of magnesium ribbon and 100 mL of household vinegar.

PROCEDURE

NOTE: Consult safety information and obtain teacher approval before beginning the experiment. Design an experiment to determine the effect of temperature on the reaction time of magnesium and vinegar.

SAFETY INFORMATION

1. Heat or cool the vinegar in hot or cold water baths; do not heat the vinegar directly.
2. Keep the temperature of the vinegar between 2°C and 85°C.
3. Do not let the magnesium ribbon contact an open flame.
4. After reviewing your procedure, the instructor will discuss any safety precautions that are specific to your experiment.

QUESTIONS FOR FURTHER THOUGHT

1. What is an example of a reaction with a fast rate? a slow rate?
2. If a scientist wanted to perform the same reaction as the one used in this experiment and wanted to increase the rate of the reaction as much as possible, what are some factors that could be changed? Explain why each of the factors would increase the rate of the reaction.

23 The Effect of Temperature on the Reaction Time of Magnesium and Vinegar

TEACHER'S NOTES

In this experiment, students investigate the effect of temperature on the time required for a reaction between magnesium and vinegar to go to completion. This experiment can be performed when students are studying kinetics and factors that affect the rate of reactions, acids, heat, and energy or single replacement reactions.

OPTIONS

1. Some students could do the same reaction using 1 M HCl instead of vinegar and compare the results.

2. Students could also investigate the effect of concentration on this reaction. If you would like students to develop an experiment to test this aspect of the reaction, use various concentrations of HCl such as 1, 2, and 3 M. (Diluting household vinegar can cause the reaction times to become so lengthy that the students cannot finish the lab in a reasonable time period.)

NOTES

1. If students run one trial at a time, the lab can take as long as 60 minutes (using 1 cm magnesium per trial) (see data in the sample lab report). You may want to encourage each team member to be responsible for one or two trials so the experiment can be completed in one lab period, or you can substitute 1 M HCl for the vinegar to increase the reaction rate.

2. Students may be concerned about the volume of vinegar required to completely react with the magnesium ribbon. You may want to recommend a minimum volume of 10 mL (per 1 cm Mg) or provide the students with an extra sample so they can run a trial before developing a procedure.

TIME

The amount of time needed for students to design and execute their own experiment can vary depending on the level and creativity of the class, their familiarity with designing their own investigations, and the complexity of the procedure they develop.

Average Time for This Experiment

Time needed to design the experiment: 15–20 minutes

Time needed to run the experiment: 45 minutes (with different team members performing separate trials)

TEAMS

Teams of two to three work well.

MATERIALS (PER TEAM)

Require the use of goggles, gloves, and aprons/lab coats. The following supplies should be readily available for the students:

- Magnesium ribbon (5 cm)
- Household vinegar (100 mL)
- Heat source
- Ice
- Test tubes (~6)
- Test tube rack
- Test tube clamp
- Thermometers (2)
- Stopwatch
- 10-mL graduated cylinder
- 100-mL or 250-mL beakers
- Scissors (to cut the Mg ribbon)
- Ruler

SAMPLE PROCEDURES

Many students will develop an experiment in which they divide the magnesium ribbon into samples of equal size and add each sample to a test tube of vinegar that has been preheated or precooled (in a water bath) to a selected temperature. Students record the length of time needed for the magnesium to completely react.

SAFETY PRECAUTIONS

NOTE: Also refer to the safety information given with the experiment. Discuss with the students any safety precautions that are specific to the procedure they develop. Material Safety Data Sheets for each substance used should be reviewed and made available to the students.

COMMON MISCONCEPTIONS, PROCEDURAL AND CALCULATION ERRORS

1. Students might vary the volume of vinegar or length of magnesium used in each trial.
2. Students might vary both the volume of vinegar and length of magnesium used in each trial.
3. Students might only perform two trials, which will produce a linear relationship if the students are analyzing the data graphically.
4. Students might have difficulty determining when the reaction has reached completion.

LAB REPORT

There are several alternatives to writing a formal lab report given in the introduction. If students are writing their own lab report it should include:

I. Title
II. Purpose
III. Materials
IV. Procedure
V. Data
VI. Calculations and/or graph
VII. Conclusion (including a discussion of sources of error)
VIII. Answers to Questions (optional)

ANSWERS TO THE QUESTIONS

1. Student answers will vary. Reactions with fast reaction rates might include double replacement reactions and the formation of a precipitate, active metals in water, an explosion or bomb, and so on. Reactions with slow reaction rates would include rusting, tarnishing, the reaction of acid rain on marble and other materials, and so on.
2. The rate of the reaction used in the experiment could be increased by raising the temperature of the acid, cutting the magnesium into smaller pieces, or adding a catalyst. Raising the temperature of the acid increases the kinetic energy of the particles and therefore the number of effective collisions that will occur per unit of time. Cutting the magnesium into smaller pieces increases the number of collisions that occur between the magnesium and acid, and a catalyst would produce a new reaction mechanism with a lower activation energy.

Acknowledgments Adapted from Borgford, C.L., Summerlin, L.R. *Chemical Activities.* American Chemical Society, Washington, DC, 1988, pp. 90–91.

23 The Effect of Temperature on the Reaction Time of Magnesium and Vinegar

SAMPLE LAB REPORT

PURPOSE

To determine the relationship between temperature and the time needed for the reaction between magnesium and vinegar to be completed.

MATERIALS

Safety equipment: goggles, gloves, and apron
Magnesium ribbon (5 cm)
Household vinegar (50 mL)
Hot plate
Test tubes (5)
Test tube rack
Thermometers (2)
Stopwatches (2)
10-mL graduated cylinder
250-mL beaker
Test tube holder
Scissors
Ruler

PROCEDURE

1. Cut the magnesium ribbon into five equal pieces of 1 cm each.
2. Place 10 mL of vinegar into each of five test tubes.
3. Measure the temperature of the vinegar in one of the test tubes.
4. Place one piece of Mg into the same test tube of vinegar (from step #3) and time how long it takes for the magnesium to completely react.
5. Place about 150 mL of water in the 250-mL beaker.

6. Put the remaining four test tubes in the 250-mL beaker and begin heating the water.
7. Place a thermometer in one of the test tubes of vinegar in the beaker.
8. When the temperature of the vinegar reaches 10°C above room temperature, remove the test tube from the water (using a test tube holder) and add a piece of magnesium ribbon. Time how long it takes for the magnesium to completely react. Record the temperature of the reaction.
9. Repeat steps #7 and 8 at temperatures 20°, 30° and 40°C above room temperature.

DATA

TRIAL	TEMP. (°C)	TIME (min)
1	22	16.4
2	31	13.4
3	41	10.1
4	51	7.5
5	61	5.6

GRAPH

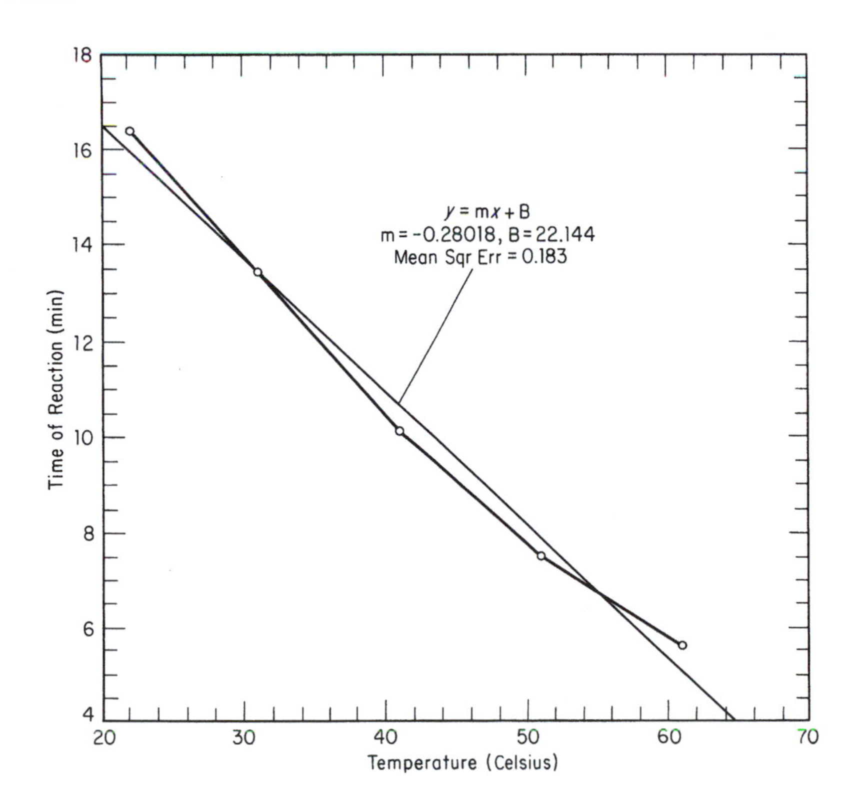

CONCLUSION

The graph suggests a linear relationship between the variables with the equation: Time (min) = −0.281 (temp.) + 22.1

It can be concluded that as the temperature of the vinegar increased, the time of the reaction decreased. This is consistent with the collision theory, which states that molecules must collide at the correct angle and with sufficient energy for a reaction to take place. The higher the temperature of the reaction, the greater the percentage of molecules that will collide with the required amount of energy.

SOURCES OF ERROR

The following is a compilation of sources of error that students may suggest:

1. If the reaction is very endothermic or exothermic, the temperature of the vinegar will change as the reaction proceeds.
2. The temperature of the vinegar did not remain constant because the test tube was removed from the water bath when the magnesium was added.
3. The magnesium strips may not have been identical in length and thickness.
4. The volume of vinegar may have been a little different in the various trials.
5. It was difficult to determine when the magnesium ribbon finished reacting.

24 Brown versus White Eggshells

THE EXPERIMENT

PURPOSE

To determine and compare the calcium carbonate content of brown and white eggshells.

BACKGROUND

Calcium carbonate ($CaCO_3$) is a component of seashells and eggshells, giving them their strength and hardness. When calcium carbonate reacts with hydrochloric acid, the products are carbon dioxide, water, and calcium chloride, according to the equation:

$$CaCO_{3(s)} + HCl_{(aq)} \rightarrow CaCl_{2(aq)} + CO_{2(g)} + H_2O_{(l)}$$

The portion of the eggshell that is not calcium carbonate does not react with the acid and remains as a solid.

In this experiment, you will be reacting samples of eggshell with 3 M HCl to compare the composition of brown and white eggshells.

PROCEDURE

NOTE: Consult safety information and obtain teacher approval before beginnning the experiment. Design an experiment to determine what percentage of an eggshell's mass is calcium carbonate, and then compare the composition of brown and white eggshells.

SAFETY INFORMATION

1. Excess hydrochloric acid may be present at the conclusion of the reaction. Take appropriate precautions.
2. Hydrochloric acid is caustic. External contact can cause severe burns. If contact does occur, wash the area immediately with water for 15 minutes and notify the instructor. Contact with the eyes can cause permanent damage. If contact with the eyes occurs, use an eyewash for 15 minutes and call for an ambulance.

3. Always wear appropriate goggles and aprons when handling this acid.
4. After reviewing your procedure, the instructor will discuss any safety precautions that are specific to your experiment.

QUESTIONS FOR FURTHER THOUGHT

1. Calcium carbonate decomposes upon heating. What are the products of the decomposition reaction?
2. Calcium carbonate is an abundant mineral that is found in several different forms. What are some of the forms of calcium carbonate?

TEACHER'S NOTES

In this experiment, students are asked to determine what percentage of an eggshell's mass is due to calcium carbonate and then compare the percentage composition of brown and white eggshells.

This experiment can be done when students are studying reactions of acids, the stoichiometry of chemical equations, or percentage composition.

OPTIONS

1. You can divide the class in half, having some teams study brown eggshells and the other teams study white eggshells.
2. Seashells that can be crushed in a mortar and pestle could also be used.

NOTES

1. Eggshells are approximately 78% calcium carbonate by mass.[1]
2. You will need to remove the inner membrane of the eggshell before using them in the experiment or have the students remove it before beginning the experiment.
3. Each team will need ½ to ¾ of an eggshell.
4. A large amount of foaming will occur when the acid is added to the eggshell. If students continue to stir the mixture, the foam will dissipate. The foaming is due to the production of carbon dioxide. If students use a 100-mL beaker, the foam might overflow; therefore, the use of 250-mL beakers (or larger) is recommended.
5. Some students will add the acid directly to the eggshell, while others will crush the eggshell in a mortar and pestle before adding the acid.
6. Students will want to know how long to let the eggshell and acid react. Students should wait until all of the foam has dissipated before moving to the next step in their procedure.
7. Students may ask if they can add a little more HCl to the eggshell to make sure that all of the calcium carbonate has reacted. This is left to your discretion.
8. Inform the students that they will have time another day to determine the mass of anything left to dry overnight.
9. You may want to withhold the HCl from the students until their procedure has been approved.

TIME

The amount of time needed for students to design and execute their own experiment can vary depending on the level and creativity of the class, their familiarity with designing their own investigations, and the complexity of the procedure they develop.

Average Time for This Experiment

Time needed to design the experiment: 20 minutes
Time needed to run the experiment: 50–80 minutes per eggshell (plus time another day to mass remaining shell)

TEAMS

Teams of two to three work well.

MATERIALS (PER TEAM)

Require the use of goggles, gloves, and aprons/lab coats. The following supplies should be readily available for the students:

- Eggshell (brown and/or white)
- 15 mL of 3.0 M HCl (dissolve 250 mL of concentrated HCl/liter of solution)
- 250-mL beakers (or larger)
- 50-mL graduated cylinder
- Balance with a precision of ± 0.01 g
- Stirring rod
- Rubber policeman
- Filter paper
- Funnel
- Mortar and pestle
- Pneumatic trough, rubber tubing, one-holed stoppers, and so on (to capture CO_2, if needed)

SAMPLE PROCEDURES

1. Many students develop a procedure in which they add the acid to the eggshell (either whole or crushed) and stir until the bubbling stops and the foam dissipates. Then the students filter the mixture and determine the mass of the remaining eggshell (after drying overnight). This will indicate the mass of eggshell that is not calcium carbonate. From that, they calculate the mass of calcium carbonate present in the shell and the percentage composition.

2. Some teams might follow the above procedure, but discard the remaining eggshell, and allow the water in the filtrate to evaporate, leaving calcium chloride behind. These students should not be permitted to boil the solution to dryness due to the presence of excess HCl. This method takes the longest to complete.

3. Some students may measure the volume of carbon dioxide gas that is produced using water displacement. Then the students use the mole ratio of the balanced equation to determine how much calcium carbonate was present. These students would need to calculate the number of moles of carbon dioxide produced using the ideal gas law formula, measure the barometric pressure, and take the vapor pressure of water into account.

4. Some students may measure the mass of the dried eggshell and acid before the reaction and compare that value to the final mass of the mixture. The difference in the two measurements represents the mass of carbon dioxide produced. Then, using the stoichiometry of the balanced equation, students calculate the mass of calcium carbonate that reacted.

SAFETY PRECAUTIONS

NOTE: Also refer to the safety information given with the experiment. Discuss with the students any safety precautions that are specific to the procedure they develop. Material Safety Data Sheets for each substance used should be reviewed and made available to the students.

Specific precautions in this lab involve the safe handling and use of hydrochloric acid. Students should be reminded that there will be excess HCl in their reaction vessel at the end of the experiment and that they should continue to be careful with the mixture.

COMMON MISCONCEPTIONS, PROCEDURAL AND CALCULATION ERRORS

1. Students may forget that the portion of the eggshell that is not $CaCO_3$ will still be present in the solid phase and can therefore be filtered.
2. Students may think that the only way to do this experiment is to measure the mass of HCl used, measure the total mass of all of the products of the reaction, and subtract to obtain the mass of the calcium carbonate that reacted.
3. Students may not pay attention to the phases of the products of the reaction when initially designing the procedure.
4. Students may think that they have to measure the mass of $CaCO_3$ directly rather than calculate its mass from other measurements.

LAB REPORT

There are several alternatives to writing a formal lab report given in the introduction. If students are writing their own lab report it should include:

I. Title
II. Purpose
III. Materials
IV. Procedure
V. Data
VI. Calculations
VII. Conclusion (with a discussion of sources of error)
VIII. Answers to Questions (optional)

ANSWERS TO THE QUESTIONS

1. Calcium carbonate decomposes to produce calcium oxide and carbon dioxide. The balanced equation for the reaction is:

   $$CaCO_{3(s)} \rightarrow CaO_{(s)} + CO_{2(g)}$$

2. Different forms of calcium carbonate include limestone (the most common form of calcium carbonate), calcite, marble, and precipitated chalk.

REFERENCE

1. Tocci, S., Viehland, C. *Chemistry: Visualizing Matter.* Holt, Rinehart, and Winston, Austin, TX, 1996, p. 760.

24 Brown versus White Eggshells

SAMPLE LAB REPORT

PURPOSE

To determine the calcium carbonate content of brown and white eggshells.

MATERIALS

Safety equipment: goggles, gloves, and apron
Brown and white eggshells
3 M HCl (15.0 mL)
250-mL beakers (2)
50-mL graduated cylinder
Stirring rod
Rubber policeman
Funnel
Filter paper
Mortar and pestle
Weighing dish
Balance

PROCEDURE

1. Obtain a sample of brown or white eggshell.
2. Crush the eggshell with a mortar and pestle.
3. Remove the crushed eggshell from the mortar and pestle and determine its mass.
4. Add the eggshell to the 250-mL beaker.
5. Add the acid to the beaker and stir until the bubbling and foaming stops.
6. Measure the mass of the filter paper.
7. Fold the filter paper and place it in the funnel.
8. Pour the mixture from the beaker into the filter paper, allowing the filtrate to pass into the other 250-mL beaker.
9. Use the rubber policeman to scrape the unreacted eggshell from the beaker into the filter paper.
10. Rinse the unreacted eggshell (in the filter paper) with water.
11. Place the filter paper where it won't be disturbed and allow it to dry overnight.
12. Repeat steps 1–11 with the eggshell of other color.
13. After drying, determine the mass of the both filter papers and unreacted eggshell.

DATA

White Eggshell

Mass of eggshell: 2.85 g
Mass of filter paper: 0.98 g
Mass of filter paper and unreacted eggshell: 1.59 g

CALCULATIONS

White Eggshell

1. Mass of unreacted eggshell = (mass of filter paper and eggshell − mass of filter paper)
= 1.59 g − 0.98 g = 0.61 g

2. Mass of calcium carbonate in original eggshell = (mass of eggshell − mass of unreacted eggshell)
= 2.85 g − 0.61 g
= 2.24 g
3. % $CaCO_3$ in the eggshell = (mass of calcium carbonate in eggshell/mass of entire eggshell)
= 2.24 g calcium carbonate/2.85 g eggshell
= 0.786 × 100 = 78.6% $CaCO_3$

DATA

Brown Eggshell

Mass of eggshell: 4.34 g
Mass of filter paper: 1.23 g
Mass of filter paper and unreacted eggshell: 3.16 g

CALCULATIONS

Brown Eggshell

1. Mass of unreacted eggshell = (mass of filter paper and unreacted eggshell − mass of filter paper)
= 3.16 g − 1.23 g = 1.93 g
2. Mass of calcium carbonate in original eggshell = (mass of eggshell − mass of unreacted eggshell)
= 4.34 g − 1.93 g
= 2.41 g
3. % $CaCO_3$ in the eggshell = (mass of calcium carbonate in eggshell/mass of entire eggshell)
= 2.41 g calcium carbonate/4.34 g eggshell
= 0.555 × 100 = 55.5% $CaCO_3$

CONCLUSION

From our calculations, white eggshells are approximately 78.6% calcium carbonate (by mass), and brown eggs are 55.5% calcium carbonate.

SOURCES OF ERROR

The following is a compilation of sources of error that students may suggest:

1. We were unable to remove all of the unreacted eggshell from the beaker.
2. We aren't sure if all of the calcium carbonate in the eggshell reacted.
3. Some of the aqueous $CaCl_2$ may have still been on the unreacted eggshell (and not rinsed through with the water), which would change the mass of the unreacted eggshell.
4. Some of the unreacted eggshell stuck to the rubber policeman when we were trying to remove it from the beaker.
5. Human error and mechanical error.

25 Comparing the Rate at Which Water Cools in Various Containers

THE EXPERIMENT

PURPOSE

To compare the rate at which water cools in various containers.

BACKGROUND

In this experiment, you will be given containers composed of different materials, such as ceramic, metal, Pyrex, and Styrofoam.

Ceramics are typically made from clays and are hardened by heating them to high temperatures. Ceramics are nonmetallic, strong, brittle, and are resistant to attack by chemicals.

Properties of metals include high electrical conductivity, malleability, and ductility. These properties are due to the ease with which electrons flow throughout the metal.

Styrofoam was the first foam plastic to be introduced and was produced by Dow Chemical Company in 1944. It is commonly used as a building insulation and for floatation devices. The common Styrofoam cup is created in a mold using expandable beads.

Pyrex is the most common brand of borosilicate glass and was developed at Corning Glass Works. It has a lower thermal expansion coefficient than other forms of glass and is therefore resistant to thermal shock, making it useful in the laboratory.

PROCEDURE

NOTE: Consult safety information and obtain teacher approval before beginning the experiment. Design an experiment to compare the rate at which heat is lost from water in each of the containers.

SAFETY INFORMATION

1. Do not directly heat the containers.
2. After reviewing your procedure, the instructor will discuss any safety precautions that are specific to your experiment.

QUESTIONS FOR FURTHER THOUGHT

1. What is the difference between heat and temperature?
2. How are heat and temperature measured?
3. Does an absorption of heat always produce a change in temperature? Explain.

25 Comparing the Rate at Which Water Cools in Various Containers

TEACHER'S NOTES

In this experiment, students will compare the rate at which water cools in containers made of various materials. This experiment can be conducted when students are studying heat, calorimetry, or properties of materials.

OPTIONS

1. Students could be asked to calculate the quantity of heat lost from the containers per unit of time by using $\Delta H = (\text{mass})(\Delta T)(\text{specific heat})$.
2. Students could be limited to one type of material but have containers of various diameters, allowing the students to study the relationship between the surface area and the rate of cooling.
3. Plastic containers able to withstand the temperature of boiling water could also be used.
4. This lab works well with temperature probes that interface with a computer. Students generally obtain graphs from which they can calculate the rate of change for each container by determining the slope of the best-fit line.

NOTES

1. The use of glass containers is not recommended.
2. The physics department of your school may have metal containers (cups) that can be used in this lab. Ceramic mugs, heavy plastic cups, and Pyrex beakers work well as the other containers.
3. You will want to have insulated gloves, oven mitts, or beaker tongs available for the students to use while pouring hot water into the various containers.
4. The lab works best if each team has its own set of containers. If students share containers, a given container may not have returned to room temperature in time for another team to begin.
5. Some students may not calculate the rate of cooling (i.e., they do not put the change in terms of time).
6. Some students may measure the diameter of each cup to determine the rate of cooling per minute per square centimeter.

TIME

The amount of time needed for students to design and execute their own experiment can vary depending on the level and creativity of the class, their familiarity with designing their own investigations, and the complexity of the procedure they develop.

Average Time for This Experiment

Time needed to design the experiment: 15 minutes
Time needed to run the experiment: 45–60 minutes

A large amount of time is spent heating the water. You can reduce the amount of time needed by having some preheated water available for the students, or suggest that students keep the water heating while doing individual trials.

TEAMS

Students may work individually or in teams of two. The size of the teams may depend on the number of containers available.

MATERIALS (PER TEAM)

Require the use of goggles, gloves, and aprons/lab coats. The following supplies should be readily available for the students:

- Various containers (ceramic mugs, metal cups, Pyrex beakers, Styrofoam cups, etc.)
- Thermometer or temperature probe
- Stopwatch
- 100-mL graduated cylinder
- Beaker tongs
- Insulated gloves
- Hot plate or other heat source
- Wire gauze
- Rulers

SAMPLE PROCEDURES

Most students develop an experiment in which they heat a given amount of water to a certain temperature and then pour it into one of the containers. Then they either wait a certain amount of time and measure the temperature of the water or time how long it takes the water to drop to a certain temperature or lose a certain number of degrees.

SAFETY PRECAUTIONS

NOTE: Also refer to the safety information given with the experiment. Discuss with the students any safety precautions that are specific to the procedure they develop. Material Safety Data Sheets for each substance used should be reviewed and made available to the students.

COMMON MISCONCEPTIONS, PROCEDURAL AND CALCULATION ERRORS

1. Students might measure the volume of water to be used before they heat it rather than before pouring it into the container. This will not account for the loss of water due to evaporation during the heating process.
2. Students may measure the temperature of the water before they pour it into the container, which will not account for the loss of heat to the air during pouring.
3. Students who use boiling water may assume the initial temperature is 100°C.
4. Some students might not calculate the *rate* of cooling (i.e., they do not calculate the change in temperature [or heat] per unit of time).

5. Some students might decide to record the change in temperature that occurs in the first 30 seconds or minute that the water is in the container and then continue with the next material. This can result in small temperature changes that are difficult to measure and make any differences in the cooling rates difficult to detect.
6. Students might not choose a volume of water that can be used in all of the available containers.
7. Students might start each trial at a different temperature rather than waiting for the water in the containers to reach a given temperature before starting.

LAB REPORT

There are several alternatives to writing a formal lab report given in the introduction. If students are writing their own lab report it should include:

I. Title
II. Purpose
III. Materials
IV. Procedure
V. Data
VI. Calculations and/or graph(s)
VII. Conclusion (including a discussion of sources of error)
VIII. Answers to Questions (optional)

ANSWERS TO THE QUESTIONS

1. Heat is the energy that is transferred between two systems of different temperature. Temperature is a measure of the average kinetic energy of the particles in the system.
2. Temperature is determined by measuring the expansion and contraction of various materials in thermometers. The total amount of heat in a system cannot be measured directly, but the change in heat can be calculated. To determine the change in heat of a system, the formula $\Delta H =$ (mass) (ΔT)(specific heat) is used.
3. No, the absorption of heat energy does not always result in a temperature change. An example is when water is changing phase. While water is melting or boiling, heat energy is being absorbed, but the temperature of the sample remains constant.

FOR FURTHER READING

Zumdahl, S.S. *Chemistry,* 3rd ed. D.C. Heath and Company, Lexington, MA, 1993, pp. 461–463.

25 Comparing the Rate at Which Water Cools in Various Containers

SAMPLE LAB REPORT

PURPOSE

To compare the rate at which water cools in various containers.

MATERIALS

- Safety equipment: goggles, gloves, and apron
- Ceramic mug
- Metal cup
- 250-mL beaker
- 400-mL beaker
- Styrofoam cup
- Stopwatch
- Insulated gloves
- Thermometer
- 100-mL graduated cylinder

PROCEDURE

1. Heat approximately 125 mL of water in the 400-mL beaker until it is about 60°C.
2. Using the insulated gloves, measure 100 mL of the heated water in the graduated cylinder and immediately pour it into the 250-mL Pyrex beaker.
3. Measure the temperature of the water in the container.
4. After 5 minutes, record the temperature of the water again.
5. Repeat steps 1–4 with the other containers.

NOTE: The students did not wait for the water to reach a given temperature before beginning data collection.

DATA

	CERAMIC MUG	METAL CUP	PYREX BEAKER	STYROFOAM CUP
Initial Temp.(°C)	54.1	54.3	55.1	53.9
Final Temp. (°C)	46.8	47.2	48.2	49.5

CALCULATIONS

CONTAINER	CHANGE IN TEMP. (°C in 5 min)	RATE OF COOLING (°C/minute)
Ceramic mug	7.3	1.5
Metal cup	7.1	1.4
Pyrex beaker	6.9	1.4
Styrofoam cup	4.4	0.88

CONCLUSION

The rate at which the water cooled varied among the containers. The Styrofoam cup lost heat at the slowest rate (0.88°C/minute), the Pyrex beaker and the metal cup had a rate of 1.4°C/minute, and the water cooled the fastest in the ceramic mug, with a rate of 1.5°C/minute.

SOURCES OF ERROR

The following is a compilation of sources of error that students may suggest:

1. The different containers had different diameters at the top and therefore a different surface area of water was exposed to the air. This could affect the rate of cooling.
2. All of the samples of water did not start at the same temperature, and the rate of cooling might vary with initial temperature.
3. The experiment was run over two lab periods, and therefore (a) the use of two thermometers with different calibrations would have affected the data, (b) the temperature of the room may have been different each day and that could affect the rate of cooling, and (c) the temperature of the cups may have been different each day and this would also affect the rate of cooling.

26 The Effectiveness of Various Antacids

THE EXPERIMENT

PURPOSE

To determine the effectiveness of various antacids.

BACKGROUND

Commercial antacids work to relieve indigestion and heartburn by neutralizing stomach acid. These products can contain one or more of a variety of substances that react with the hydrochloric acid found in gastric juice, thereby relieving symptoms. Ingredients commonly found in these products include calcium carbonate, $CaCO_3$; magnesium hydroxide, $Mg(OH)_2$; and aluminum hydroxide, $Al(OH)_3$.

In this experiment, you will be given two or more antacid products and 0.10 M HCl (to simulate stomach acid). Design an experiment to determine which of the products is the most effective in neutralizing stomach acid.

PROCEDURE

NOTE: Consult safety information and obtain teacher approval before beginning the experiment. Design an experiment to determine which of the antacid products is the most effective in neutralizing stomach acid, noting the following:

1. Record the active ingredients and recommended dosage for each antacid product.
2. Solid tablets should be crushed in a mortar and pestle with 10 mL of water before beginning.

SAFETY INFORMATION

1. Contact with hydrochloric acid can cause severe burns. If contact occurs, wash the area immediately with water for 15 minutes and notify the instructor. Contact with the eyes can cause permanent damage. If contact with the eyes occurs, use an eyewash for 15 minutes and call for an ambulance.

2. Always wear appropriate goggles and apron when handling this acid.
3. After reviewing your procedure, the instructor will discuss any safety precautions that are specific to your experiment.

QUESTIONS FOR FURTHER THOUGHT

1. Write a balanced equation for the reaction of each active ingredient with HCl.
2. Why do some people who do not suffer from indigestion regularly take antacid tablets?

26 The Effectiveness of Various Antacids

TEACHER'S NOTES

In this experiment, students are given two or three commercial antacid products and asked to determine which of the products is the most effective in neutralizing stomach acid (0.1 M HCl).

This experiment works best with students who are familiar with the use of burets, indicators, and the technique of titration, although students can perform the lab simply using pH paper or litmus paper.

OPTIONS

1. Different teams in the class can test different products and compile their results.
2. Students can test 1 tablespoon or 1 tablet of each brand or test the recommended dosage given on the label.
3. Students can bring outdated products from home and determine if they have lost their effectiveness.
4. Students can compare the tablet and liquid form of the same product.
5. Students can compare products that have the same active ingredients; perhaps choosing a name brand and a generic brand.

NOTES

1. If students decide to titrate the acid and antacid using an indicator to determine the endpoint, they will need to decide what indicator to use, or you can select one for them. Note the following when approving a procedure:

For students adding the acid to the antacid (whether they are using a buret or counting drops from an eyedropper), they should select an indicator that will change color around pH 7 or 8. You may want to show the students the chart given in Appendix 4 and let them choose the indicator. Phenolphthalein works, although it can change color after one or two drops with some products, making a comparison of the products difficult. Better choices include bromothymol blue, phenol red, and bromcresol purple. If students select an indicator that changes color in the range of pH 3–5, such as methyl orange, they may need to add 100–150 mL of acid to obtain a color change. This can extend the amount of time needed to complete the experiment as well as the volume of acid you will need to have available for the students.

For students adding the antacid to the acid (using an eyedropper rather than a buret, see note #2 below), the students should choose an indicator that changes color in the acid range. Students can again refer to Appendix 4. In this case, methyl orange works well, with a color change occuring after 20–30 drops (per tablet used) for most products.

2. If students decide to use a buret for the experiment, emphasize that the *acid* should be placed in the buret. (The antacid products are difficult to clean from the buret and stopcock.)
3. Antacid products that have a color can make it more difficult to see a

color change when doing a titration. If you purchase the products in a store, try to buy products that are white.

4. Check the labels carefully when buying the products, as many antacids contain the same active ingredients.

5. Antacids that contain calcium carbonate will fizz upon the addition of acid according to the equation:

$$CaCO_{3(s)} + HCl_{(aq)} \rightarrow CaCl_{2(aq)} + H_2O_{(l)} + CO_{2(g)}.$$

Products containing hydroxides, such as $Mg(OH)_2$ or $Al(OH)_3$, do not show the same reaction. The production of carbon dioxide in the stomach will often cause the person to burp, aiding in the release of pressure often associated with heartburn and indigestion.

6. Students should not measure the volume of the liquid antacids in a graduated cylinder. The use of spoons will aid in cleanup.

TIME

The amount of time needed for students to design and execute their own experiment can vary depending on the level and creativity of the class, their familiarity with designing their own investigations, and the complexity of the procedure they develop.

Average Time for This Experiment

Time needed to design the experiment: 20 minutes
Time needed to run the experiment: 50 minutes

TEAMS

Teams of two work well.

MATERIALS (PER TEAM)

Require the use of goggles, gloves, and aprons/lab coats. The following supplies should be readily available for the students:

- Several antacid products
- 0.1 M HCl (50 mL per team; dissolve 8.6 mL concentrated acid per liter solution)
- Mortar and pestle
- Buret
- Dropper pipet
- Buret clamp
- Ring stand
- Stirring rod
- Graduated cylinders (10-mL, 50-mL)
- Appropriate indicators
- pH paper
- Litmus paper
- Distilled water
- Beakers (various sizes)
- Rubber policeman
- Teaspoon or tablespoon

SAMPLE PROCEDURES

1. Students may add a given amount of acid (e.g., 10 mL) to a given amount of antacid (perhaps 1 tablet, 1 tablespoon, or the recommended dose) and check the resulting pH with pH paper. Students compare the pH of the

mixtures to determine which antacid did the best job of neutralizing the given volume of acid.

2. Students may add a given volume of acid (e.g., 1 mL) to a given amount of antacid (perhaps 1 tablet, 1 tablespoon, or the recommended dose) and check the pH, after the addition of acid, with blue litmus paper. The students continue to add acid and measure the volume needed for the blue litmus paper to turn red.
3. Students may use method #2 above, only adding the antacid product to the acid and using red litmus paper.
4. Some students may simply take the pH of the product (after grinding the tablets in water) and decide that the product with the highest pH is the most effective product.
5. Students may add the antacid product to a given volume of acid (containing an indicator), using a dropper pipette (see #1 in the Notes), and record the number of drops needed to produce a color change in the acid. Students determine that the product that required the smallest volume to produce the color change is the most effective product.
6. Students may add acid to the recommended dose of antacid (containing an indicator) using a buret or dropper pipette (see #1 in the Notes) and record the number of drops, or volume of acid, required to produce a color change. Students then use the formula $V_aN_a = V_bN_b$ to determine the normality of the antacid product or assume that the product that required the smallest volume to produce the color change is the most effective product.

SAFETY PRECAUTIONS

NOTE: Also refer to the safety information given with the experiment. Discuss with the students any safety precautions that are specific to the procedure they develop. Material Safety Data Sheets for each substance used should be reviewed and made available to the students. A specific precaution in this lab involves the handling of the hydrochloric acid.

COMMON MISCONCEPTIONS, PROCEDURAL AND CALCULATION ERRORS

1. Students may think that indicators are different shades of color at different pH.
2. Students may not consider which indicator to use in their titration (see Note #1).

LAB REPORT

There are several alternatives to writing a formal lab report given in the introduction. If students are writing their own lab report it should include:

I. Title
II. Purpose
III. Materials
IV. Procedure
V. Data

VI. Calculations
VII. Conclusion (including a discussion of sources of error)
VIII. Answers to Questions (optional)

ANSWERS TO THE QUESTIONS

1. The reaction of some of the active ingredients found in antacid products are shown below:
   $CaCO_3 + HCl \rightarrow CaCl_2 + H_2O + CO_2$
   $Al(OH)_3 + 3HCl \rightarrow AlCl_3 + 3H_2O$
   $Mg(OH)_2 + 2HCl \rightarrow MgCl_2 + 2H_2O$
   $Ca(OH)_2 + 2HCl \rightarrow CaCl_2 + 2H_2O$
   $MgCO_3 + 2HCl \rightarrow MgCl_2 + CO_2 + H_2O$
2. Some people take antacid tablets regularly as a dietary supplement. For example, some antacid products contain calcium, which people take to maintain bone strength and prevent osteoporosis.

26 The Effectiveness of Various Antacids

SAMPLE LAB REPORT

PURPOSE

To determine the effectiveness of various antacids.

MATERIALS

- Safety equipment: goggles, gloves, and apron
- Antacid products (1 liquid, 2 tablets)
- 0.1 M HCl
- Bromthymol blue indicator
- Buret
- Buret clamp
- Ring stand
- Mortar and pestle
- Distilled water
- Stirring rod and rubber policeman
- 250-mL beaker

PROCEDURE

1. Rinse the buret with distilled water.
2. Add 50 mL of 0.1 M HCl to the buret.
3. Obtain one of the antacid products to be used. If it is a tablet, crush it in a mortar and pestle with 10 mL of water and then place it in the 250 mL beaker. If it is a liquid antacid, put 1 tablespoon of the liquid into the 250-mL beaker.
4. Add 3–4 drops of bromthymol blue to the antacid in the beaker.

5. Slowly add HCl from the buret into the beaker until the mixture changes color. Stir the mixture as the acid is being added. Record the volume of acid in the buret.
6. Repeat steps 3–5 with the other antacid products.
7. Record the name of the products and their active ingredients.

PRODUCT	FORM	QUANTITY OF ACTIVE INGREDIENT
1	Tablet	500 mg $CaCO_3$ per tablet
2	Tablet	550 mg $CaCO_3$ and 110 mg $Mg(OH)_2$ per tablet
3	Liquid	200 mg $Mg(OH)_2$ and 225 mg $Al(OH)_3$ per 5 mL tsp.

PRODUCT	VOLUME OF ACID ADDED (mL)	OBSERVATIONS
1	3.2	Fizzing
2	2.5	—
3	1.2	—

DATA

CONCLUSION

According to our data, product #1 is the most effective product because it took the largest volume of acid to neutralize it, making this product the strongest base. Product #2 is the second best product, and product #3 is the worst.

SOURCES OF ERROR

The following is a compilation of sources of error that students may suggest:

1. It was difficult to tell exactly when the indicator changed color.
2. We wanted to use the recommended dose for each product, but the recommended dosage varied for some antacids. For example, some bottles would say that the recommended dosage is 2–4 tablets or 3–5 teaspoons. We decided to use 1 tablet and 1 teaspoon, but that is not consistent with what a person might really take.
3. Some of the tablet may have remained in the mortar and pestle.
4. We aren't sure if all of the tablet dissolved.
5. We might have added the acid too fast, not allowing the reaction or color change enough time to take place. This would have caused us to add more acid than needed.

27 The Juice Lab

THE EXPERIMENT

PURPOSE

To determine the concentration at which various juices (such as grape juice or cranberry juice) lose their color and smell.

BACKGROUND

A common practice in many laboratories is to only purchase concentrated forms of solutions. When a more dilute solution is required, water (or another solvent) is added to a given amount of the concentrated form to produce the concentration required. A formula that relates the initial concentration of the solution to the final concentration is:

$$C_i V_i = C_f V_f$$

where, C_i = initial concentration (in any unit); V_i = initial volume of the concentrated solution; C_f = final concentration of the diluted solution; and V_f = final volume of the diluted solution.

In this experiment, you will be given a sample of one or more types of juice and asked to determine the concentration at which the juice loses its color and smell. You are to treat color and smell separately (i.e., you are not determining the point at which both color and smell are lost).

Assume the initial concentration of the juice to be 100%.

PROCEDURE

NOTE: Consult safety information and obtain teacher approval before beginning the experiment. Determine the dilution at which various juices (such as grape juice or cranberry juice) lose their color and smell.

SAFETY INFORMATION

After reviewing your procedure, the instructor will discuss any safety precautions that are specific to your experiment.

QUESTIONS FOR FURTHER THOUGHT

1. A 20 mL sample of 3.0 M HCl is diluted to a final volume of 100 mL. What is the concentration of the diluted acid?
2. A 50 mL sample of 6.0 M HCl is found in the lab. The student needs a 2.0 M solution. How much water should the student add to the solution to prepare a 2.0 M solution?

27 The Juice Lab

TEACHER'S NOTES

This lab is intended to introduce students to the techniques and calculations involved in the dilution of solutions. You can also introduce the concept of serial dilutions if desired (see Notes #1 and 2). The lab can be used when the students are studying solutions and units of concentration.

OPTIONS

1. Students can compare brands of grape juice and/or cranberry juice.
2. Students can compare bottled and frozen forms of the same juice(s).
3. Students can study other types of juices (e.g., pineapple, orange without pulp, or apple juice).
4. You can limit the size of the containers available to the students (see Notes below).
5. You can have the students predict if the juice will lose its color or smell first, or which juice will lose its color and smell first.

NOTES

1. If students decide to start with 10 mL of juice and keep adding water until the color and smell are gone, the final volume of the solution can be approximately 1.5 liters (for grape juice and cranberry juice). If you approve this procedure, you need to either:

   A. Have enough large containers available for the entire class,
   B. Tell the students that there aren't any large containers, at which point the students may decide to either take a sample of their diluted solution and continue diluting it in another container, or start over with a smaller initial volume of juice, or
   C. Introduce the concept of serial dilutions (perhaps having pipets available).

2. You can point out to the students that because the product of these two variables is a constant, the variables themselves (volume and concentration) are inversely proportional.
3. Emphasize to the students that V_f in the equation is the final volume of the mixture, not the volume of water added during the dilution.
4. If students will be using volumetric pipettes for the first time, you may want to give them some time to practice with distilled water before beginning the lab.

TIME

The amount of time needed for students to design and execute their own experiment can vary depending on the level and creativity of the class, their familiarity with designing their own investigations, and the complexity of the procedure they develop.

Average Time for This Experiment

Time needed to design the experiment: 15 minutes
Time needed to run the experiment: 30–45 minutes

TEAMS

The experiment may be performed individually or by teams of two.

MATERIALS (PER TEAM)

Require the ues of goggles, gloves, and aprons/lab coats. The following supplies should be readily available for the students:

- Juice (1–15 mL)
- Large containers (see Note #1 above)
- Pipettes (check with the biology department of the school)
- Dropper pipettes
- Beakers and flasks (various sizes including 1-L and 2-L if available)
- Graduated cylinders (10-mL, 50-mL)

SAMPLE PROCEDURES

1. Students might take a given amount of juice (1–10 mL) and continue to add a certain volume of water until the smell and color are gone. (Students check for smell and color after each subsequent addition of water.) Students may be able to perform the entire procedure in one container (600–1000 mL beaker) or at some point will be forced to remove some diluted solution, transfer it to another container, and continue diluting in the second container.
2. Students use a pipette to obtain a certain amount of juice (e.g., 1 mL) and dilute it to a certain volume. Then, they take 1.0 mL of the diluted solution and use a pipette to transfer it to another container and dilute to the same volume. Students continue the serial dilution until the color and smell are gone.
3. Some students may use the previous technique of serial dilutions, but repeat the final dilution with smaller quantities of water to be as precise as possible.

SAFETY PRECAUTIONS

Discuss with the students any safety precautions that are specific to the procedure they develop. Material Safety Data Sheets for each substance used should be reviewed and made available to the students.

COMMON MISCONCEPTIONS, PROCEDURAL AND CALCULATION ERRORS

A potential calculation error is that students use the volume of water added to the juice as V_f rather than the final volume of the solution (volume of water added + initial volume of juice).

LAB REPORT

There are several alternatives to writing a formal lab report given in the introduction. If students are writing their own lab report it should include:

I. Title
II. Purpose
III. Materials
IV. Procedure
V. Data
VI. Calculations
VII. Conclusion (including a discussion of sources of error)
VIII. Answers to Questions (optional)

ANSWERS TO THE QUESTIONS

1. $(20.0 \text{ mL})(3.0 \text{ M}) = (100.0 \text{ mL})(x)$
   $x = 0.60$ M HCl solution
2. $(50.0 \text{ mL})(6.0 \text{ M}) = (x)(2.0 \text{ M})$
   $x = 150$ mL = volume of final solution
   volume of water that
   must be added = 150 mL − 50.0 mL = 100 mL water

Acknowledgment Thanks to Arthur B. Ellis (University of Wisconsin-Madison) for recommending this activity and to Richard Zare of Stanford University for the concept behind this lab.

FOR FURTHER READING

Kamin, L. *The American Biology Teacher.* 1996, 58, pp. 296–298.

27 The Juice Lab

SAMPLE LAB REPORT

PURPOSE

To determine the dilution at which grape juice and cranberry juice lose their color and smell.

MATERIALS

Safety equipment: goggles, gloves, and apron
Grape juice
Cranberry juice
1-mL volumetric pipette
50-mL graduated cylinder
2-L flask
250-mL beakers (2)

PROCEDURE 1

1. Place 10 mL of grape juice in the 2-L flask.
2. Add 20 mL of water and determine if the diluted juice still has an odor or color.
3. Add another 20 mL of water and again check to see if the diluted juice has an odor or color.
4. Continue adding water until the juice loses its color and smell. Record the volume of water added to the juice.
5. Repeat with the cranberry juice.

PROCEDURE 2

1. Using a pipet, put 1 mL of grape juice into a 250-mL beaker.
2. Add 1 mL of water to produce a final volume of 2 mL.
3. Test the diluted juice for color and smell.
4. If the juice still has color or smell, remove 1 mL of the diluted solution and place it in another 250-mL beaker. Add 1 mL of water. Test for color and smell.
5. Continue repeating step #4 until the juice has lost its color and smell.
6. Repeat the entire process using cranberry juice.

DATA

Procedure 1

Grape juice: Volume of water added to lose smell: 400 mL
Volume of water added to lose color: 1210 mL
Cranberry juice: Volume of water added to lose smell: 320 mL
Volume of water added to lose color: 1620 mL

Procedure 2

After the fifth dilution, the grape juice lost its color.
After the sixth dilution, the grape juice lost its smell.
After the third dilution, the cranberry juice lost its color.
After the fifth dilution, the cranberry juice lost its smell.

CALCULATIONS

Procedure 1

Concentration of grape juice when it lost its smell: $C_iV_i = C_fV_f$
$(100\%)(10.0 \text{ mL}) = (x)(400 \text{ mL})$
$x = 2.50\%$
Concentration of grape juice when it lost its color:
$(100\%)(10.0 \text{ mL}) = (x)(1210 \text{ mL})$
$x = 0.826\%$
Concentration of cranberry juice when it lost its smell:
$(100\%)(10.0 \text{ mL}) = (x)(320 \text{ mL})$
$x = 3.13\%$
Concentration of cranberry juice when it lost its color:
$(100\%)(10.0 \text{ mL}) = (x)(1620 \text{ mL})$
$x = 0.617\%$

Procedure 2

Grape juice dilutions:
1st dilution: (100%)(1.0 mL) = (x)(2.0 mL)
x = 50%
2nd dilution: (50%)(1.0 mL) = (x)(2.0 mL)
x = 25%
3rd dilution: (25%)(1.0 mL) = (x)(2.0 mL)
x = 13%
4th dilution: (13%)(1.0 mL) = (x)(2.0 mL)
x = 6.5%
5th dilution: (6.5%)(1.0 mL) = (x)(2.0 mL)
x = 3.3% (lost color)
6th dilution: (3.3%)(1.0 mL) = (x)(2.0 mL)
x = 1.7% (lost smell)

Cranberry juice dilutions:
1st dilution: (100%)(1.0 mL) = (x)(2.0 mL)
x = 50%
2nd dilution: (50%)(1.0 mL) = (x)(2.0 mL)
x = 25%
3rd dilution: (25%)(1.0 mL) = (x)(2.0 mL)
x = 13% (lost color)
4th dilution: (13%)(1.0 mL) = (x)(2.0 mL)
x = 6.5%
5th dilution: (6.5%)(1.0 mL) = (x)(2.0 mL)
x = 3.3% (lost smell)

CONCLUSION

The concentration at which the grape juice lost its color and smell varied with the two methods used. It is our opinion that procedure #1 was the most accurate because we added the water in smaller amounts. Both the grape juice and cranberry juice lost their color before they lost their smell.

SOURCES OF ERROR

The following is a compilation of sources of error that students may suggest:

1. It was hard to tell exactly when the color and smell were gone.
2. Different people can detect odors to different extents.
3. If someone had a cold, it would be harder to tell when the smell was gone.
4. It was difficult to obtain exactly 1.0 mL of juice using the pipette.
5. We assumed the bottled juice to be 100% juice, which may not be true.
6. The tap water we used may have had a scent that masked the smell of the juice.

28 The Effect of Temperature on the Rate of Diffusion

THE EXPERIMENT

PURPOSE

To determine the effect of temperature on the rate of diffusion in water.

BACKGROUND

Diffusion is the movement of molecules from an area of high concentration to an area of low concentration. The rate of diffusion is affected by temperature, pressure, the phase of the material, and the strength of the attractive forces between the particles.

In this experiment you will be studying the effect of temperature on the rate at which a liquid (such as ink or food coloring) diffuses through water.

PROCEDURE

NOTE: Consult safety information and obtain teacher approval before beginning the experiment. Design an experiment to determine the effect of temperature on the rate at which a liquid diffuses through water.

SAFETY INFORMATION

1. Regulate the temperature of the water rather than the temperature of the ink or food coloring.
2. Both ink and food coloring can stain clothing. Take appropriate precautions.
3. After reviewing your procedure, the instructor will discuss any safety precautions that are specific to your experiment.

QUESTIONS FOR FURTHER THOUGHT

1. Compare and contrast diffusion and effusion.
2. Compare the rate of diffusion of gases, liquids, and solids.
3. What relationship is given by Graham's Law regarding the rate of diffusion of gases?

28 The Effect of Temperature on the Rate of Diffusion

TEACHER'S NOTES

In this experiment, students are given a liquid such as soluble ink or food coloring and are asked to determine the effect of temperature on the rate at which the liquid diffuses in water. This experiment can be used when students are studying the kinetic theory or Graham's law.

OPTIONS

1. Students can compare the rate of diffusion of different colors of food coloring or of food coloring and ink.
2. Students can compare the rate of diffusion of the food coloring in water and in vegetable oil.

NOTES

1. Students may need to spend some time working with the ink (or food coloring) and observing the diffusion process before deciding on a procedure. Students may also need to experiment with various types of containers to determine which one to use.
2. Have a variety of containers available for the students including beakers, flasks, test tubes, gas measuring tubes, graduated cylinders, petri dishes, and so on.
3. Water-soluble black ink is available through most chemical supply companies. The type of ink used in chromatography experiments works well.
4. Students cooling the water with ice may want to remove the ice before beginning the experiment.
5. Students may either heat water in a beaker and then pour the warmed water into the container they are using or heat test tubes of water in a hot water bath.

TIME

The amount of time needed for students to design and execute their own experiment can vary depending on the level and creativity of the class, their familiarity with designing their own investigations, and the complexity of the procedure they develop.

Average Time for This Experiment

Time needed to design the experiment: 30 minutes
Time needed to run the experiment: 50 minutes

TEAMS

Teams of two work well.

MATERIALS (PER TEAM)

Require the use of goggles, gloves, and aprons/lab coats. The following supplies should be readily available for the students:

- Variety of glassware (see Note #2)
- Heat source
- Ice
- Soluble ink or food coloring
- Thermometer
- Stopwatch
- Insulated gloves or beaker tongs
- Dropper pipette

SAMPLE PROCEDURES

1. Students might place a drop of ink (or food coloring) at the surface of water in a tall container such as a graduated cylinder or gas-measuring tube and time how long it takes for some ink to reach the bottom of the container. Students using this method often obtain results that are contrary to their prediction. In the cold water, the drop of ink will sink quickly, reaching the bottom in a short period of time. While in the hot water, the drop of ink will diffuse very quickly at the surface and take a long time to reach the bottom. These students may decide to repeat the experiment using a shorter container such as a test tube.
2. Students realizing that gravity is a factor may decide to have the ink (or food coloring) diffuse horizontally through the water. These students may time how long it takes for the ink to diffuse across a petri dish, or across a given distance (e.g., 3 mL) of a gas-measuring tube that is stoppered and held horizontally.
3. Students may add a given amount of ink (or food coloring) to the water and time how long it takes for the ink to completely diffuse in the water (i.e., they will determine the point at which the ink or food coloring appears to be uniformly dispersed throughout the water).
4. Students may consider the rate of diffusion to be the time needed to diffuse, while other students will calculate rate in terms of distance traveled per unit of time or volume traveled per time. (For example, students working with a 50-mL graduated cylinder may measure the rate of diffusion in mL/second.)

SAFETY PRECAUTIONS

NOTE: Also refer to the safety information given with the experiment. Discuss with the students any safety precautions that are specific to the procedure they develop. Material Safety Data Sheets for each substance used should be reviewed and made available to the students.

COMMON MISCONCEPTIONS, PROCEDURAL AND CALCULATION ERRORS

1. Students may not understand how to measure the rate of diffusion.
2. Students may think they should heat the ink (or food coloring) rather than the water.

3. Students may use very small volumes of water which can make the rate of diffusion difficult to measure.
4. Students may not put the rate of diffusion in terms of time.

LAB REPORT

There are several alternatives to writing a formal lab report given in the introduction. If students are writing their own lab report it should include:

I. Title
II. Purpose
III. Materials
IV. Procedure
V. Data
VI. Calculations (if needed)
VII. Conclusion (including a discussion of sources of error)
VIII. Answers to Questions (optional)

ANSWERS TO THE QUESTIONS

1. Diffusion and effusion both involve molecules moving from one area to another. Diffusion is the mixing of two substances; effusion is the movement of atoms or molecules through a small opening into an evacuated container.
2. The rate at which substances diffuse is affected by the strength of the attractive forces between the particles and their ability to move relative to each other. Because solids have strong attractive forces between their particles and have only vibrational motion, the rate of diffusion is slow. In gases, the molecules are far apart, have little attraction for each other, and have translation motion. This allows gases to diffuse quickly compared to solids. Liquids diffuse at a rate in between that of solids and gases. This is due to the fact that the molecules in a liquid have some translation motion and yet the attractive forces are stronger than in the gaseous phase.
3. Graham's law states that the rate at which two gases diffuse is inversely proportional to the square root of their masses (or densities).

FOR FURTHER READING

Ellis, A.B. et. al., *Teaching General Chemistry: A Materials Science Companion.* American Chemical Society, Washington, DC, 1993, pp. 447–453.

28 The Effect of Temperature on the Rate of Diffusion

SAMPLE LAB REPORT

PURPOSE

To determine the effect of temperature on the rate of diffusion in water.

MATERIALS

- Safety equipment: goggles, gloves, and apron
- Dropper pipette
- Stopwatch
- Ink
- Thermometer
- Test tubes
- 50-mL graduated cylinder
- Hot plate

PROCEDURE

1. Pour 20 mL of water into a test tube.
2. Add 1 drop of ink to the surface of the water and time how long it takes for the ink to reach the bottom of the test tube.
3. Pour another 20 mL of water into a clean test tube and place the test tube in a hot water bath.
4. When the water in the test tube reaches approximately 40°C, record the temperature and add a drop of ink to the surface of the water. Time how long it takes for the ink to reach the bottom of the test tube.
5. Repeat steps #3 and 4 at approximately 50°C and 60°C.

DATA

TEMPERATURE (°C)	TIME (seconds)
23.2	64
40.4	10
50.6	7
60.1	6

CONCLUSION

At higher temperatures, the rate of diffusion of ink increased. This is probably due to the fact that the molecules of water have more kinetic energy at high temperatures and therefore the molecules of water and ink collide more frequently and with more energy, increasing the rate at which the ink molecules are able to move through the water molecules.

SOURCES OF ERROR

The following is a compilation of sources of error that students may suggest:

1. Gravity may have been affecting the rate at which the ink moved to the bottom of the test tube.
2. Not all drops of ink have the same volume.
3. The temperature of the water changed after removing it from the hot water bath.
4. We had difficulty stopping the stopwatch at the precise moment the ink touched the bottom of the test tube.

29 Determining the Formula of a Hydrate

THE EXPERIMENT

PURPOSE

To determine the formula of a hydrate.

BACKGROUND

Hydrates are crystalline substances that contain water chemically combined in a definite ratio. The formula of a hydrate is written to show the number of water molecules contained in the crystal for every formula unit of salt. For example, the formula $CaCl_2 \cdot 2H_2O$ indicates that for every formula unit of calcium chloride, there are two molecules of water. This also represents a mole ratio between the calcium chloride and water molecules (i.e., for every mole of calcium chloride, there are two moles of water.

When hydrates are heated, the water in the crystal is driven off in the form of steam. The solid that remains is said to be anhydrous. For some crystals, the hydrate and anhydrous forms of the substance are different colors.

For this experiment, you will be given a sample of copper sulfate hydrate, a blue crystal. During the lab, the hydrate will be heated, leaving gray anhydrous copper sulfate crystals. The equation is:

$$\underset{\text{(Blue)}}{CuSO_4 \cdot xH_2O_{(s)}} \xrightarrow{\Delta} \underset{\text{(Gray)}}{CuSO_{4(s)}} + xH_2O_{(g)}$$

You are to determine the formula of the hydrate (i.e., determine the value of x in the formula $CuSO_4 \cdot xH_2O$.

PROCEDURE

NOTE: Consult safety information and obtain teacher approval before beginnning the experiment. Design an experiment to determine the formula of copper sulfate hydrate.

SAFETY INFORMATION

1. Heat the empty evaporating dish for 3 minutes before beginning the experiment to remove any water that may have been

absorbed by the porcelain. Allow the dish to cool 2 minutes before measuring its mass or adding the crystals.
2. Stir the crystals gently with the stirring rod while heating.
3. Hold the evaporating dish with tongs.
4. Heat slowly to avoid spattering and popping.
5. Reduce the heat if the edges of the crystals begin to turn brown.
6. Do not place a hot evaporating dish directly on a balance.
7. Check with the instructor regarding the disposal of the anhydrous salt.
8. Copper sulfate hydrate is toxic by ingestion and inhalation. It is a skin and body tissue irritant. In the event of external contact, flush with water for 15 minutes and notify the instructor.
9. After reviewing your procedure, the instructor will discuss any safety precautions that are specific to your experiment.

QUESTIONS FOR FURTHER THOUGHT

1. Some hydrates are efflorescent. What does this mean?
2. How are hydrates prepared from the anhydrous form of the substance?
3. What percentage of the mass of a sample of $CaCl_2 \cdot 2H_2O$ is water?

29 Determining the Formula of a Hydrate

TEACHER'S NOTES

In this experiment, students are asked to design an experiment to determine the formula of a hydrate. The lab can be conducted when the students are studying the stoichiometry of equations or the properties of solids and crystals. An understanding of moles and the ability to convert from grams to moles is required for students to complete this experiment.

OPTIONS

1. Provide different teams with various masses of hydrate to illustrate that the mole ratio is not dependent on initial mass.
2. Ask students to calculate the percentage composition of the hydrate.

NOTES

1. Use the fine crystalline form of $CuSO_4 \cdot 5H_2O$. The large crystals or chunks are difficult to dehydrate.
2. Teams should use approximately 2–3 g of hydrate.
3. Hot plates, rather than Bunsen burners, are recommended as they produce less spattering.
4. Students are told in the safety information to heat the empty evaporating dish before beginning the experiment to drive out any water that may have been absorbed by the porcelain. This produces better data but is not absolutely necessary. If the students are limited in their time to complete the experiment, you may omit this step.
5. When approving procedures, note if students have clearly indicated what mass measurements they will be recording.
6. Students tend to have more trouble with the calculations involved in this lab than the execution of the experiment. Depending on the level of the class, you can provide more information regarding how to calculate the formula of the hydrate before the students begin.

TIME

The amount of time needed for students to design and execute their own experiment can vary depending on the level and creativity of the class, their familiarity with designing their own investigations, and the complexity of the procedure they develop.

Average Time for This Experiment

Time needed to design the experiment: 15–20 minutes
Time needed to run the experiment: 40 minutes

TEAMS

Teams of two to three work well.

MATERIALS (PER TEAM)

Require the use of goggles, gloves, and aprons/lab coats. The following supplies should be readily available for the students:

- Copper sulfate pentahydrate ($CuSO_4 \cdot 5H_2O$; 2–3 g of fine crystal)
- Evaporating dish
- Heat source
- Stirring rod
- Tongs
- Wire gauze
- Balance

SAMPLE PROCEDURES

(See sample lab report.)

SAFETY PRECAUTIONS

NOTE: Also refer to the safety information given with the experiment. Discuss with the students any safety precautions that are specific to the procedure they develop. Material Safety Data Sheets for each substance used should be reviewed and made available to the students. Students should be especially careful if they are working over an open flame. When using a Bunsen burner, students should hold the burner in their hand and gently wave the flame at the dish, occasionally stirring the crystals.

COMMON MISCONCEPTIONS, PROCEDURAL AND CALCULATION ERRORS

1. Students may use the mass ratio of copper sulfate and water rather than the mole ratio to determine the formula of the hydrate.
2. Students may compare the moles of the *hydrate* to the moles of water rather than the moles of *anhydrous* copper sulfate to the moles of water.

LAB REPORT

There are several alternatives to writing a formal lab report given in the introduction. If students are writing their own lab report it should include:

I. Title
II. Purpose
III. Materials
IV. Procedure
V. Data
VI. Calculations
VII. Conclusion (including a discussion of sources of error)
VIII. Answers to Questions (optional)

ANSWERS TO THE QUESTIONS

1. Hydrates that are efflorescent spontaneously lose the water of hydration at room temperature (i.e., they do not need to be heated).
2. The hydrate can be formed by preparing a solution of the anhydrous crystal and then allowing the water to evaporate.

3.

$$1 \text{ mol } CaCl_2 = 110.98 \text{ g}$$
$$2 \text{ mol } H_2O = 36.00 \text{ g}$$
$$\text{Formula weight of } CaCl_2 \cdot 2H_2O = 146.98 \text{ g}$$
$$\% \text{ of water in hydrate} = 36.00 \text{ g } H_2O / 146.98 \text{ g } CaCl_2 \cdot 2H_2O$$
$$= 0.2449 \times 100 = 24.49\%$$

29 Determining the Formula of a Hydrate

SAMPLE LAB REPORT

PURPOSE

To determine the formula of a hydrate.

MATERIALS

- Safety equipment: goggles, gloves, and apron
- Copper sulfate hydrate
- Hot plate
- Stirring rod
- Tongs
- Balance
- Wire gauze

PROCEDURE

1. Heat the empty evaporating dish on the hot plate for 3 minutes. Turn off the hot plate and set the dish on the wire gauze.
2. Allow the dish to cool and determine its mass.
3. Add the copper sulfate hydrate to the dish and determine the mass of the dish and crystals.
4. Heat the dish and crystals slowly on the hot plate, stirring the crystals gently with the stirring rod until all of the blue color is gone.
5. Remove the dish from the hot plate and set it on the wire gauze to cool.
6. Measure the mass of the anhydrous crystals and evaporating dish.

DATA

Mass of empty evaporating dish: 44.55 g
Mass of evaporating dish and hydrate: 47.31 g
Mass of evaporating dish and anhydrous salt: 46.28 g

CALCULATIONS

$$\text{Mass of hydrate} = \text{mass of dish with hydrate} - \text{mass of empty dish}$$
$$= 47.31 \text{ g} - 44.55 \text{ g}$$
$$= 2.76 \text{ g}$$

$$\text{Mass of anhydrous } CuSO_4 = \text{mass of dish and anhydrous salt} - \text{mass of empty dish}$$
$$= 46.28 \text{ g} - 44.55 \text{ g}$$
$$= 1.73 \text{ g}$$

Mass of water driven from hydrate = mass of hydrate − mass of anhydrous salt
= 2.76 g − 1.73 g
= 1.03 g

Moles of $CuSO_4$ contained in hydrate $= 1.73\ g \times \frac{1\ mol}{159.60\ g} = 0.0108\ mol$ $CuSO_4$

Moles of H_2O contained in hydrate $= 1.03\ g \times \frac{1\ mol}{18.00\ g} = 0.0572\ mol\ H_2O$

Ratio of moles of water to moles of $CuSO_4$ $= \frac{0.0572\ mol\ H_2O}{0.0108\ mol\ CuSO_4} = 5.30\ (\sim 5)$

CONCLUSION

The data showed that for every mole of $CuSO_4$, there were approximately 5 moles of water. The formula of the hydrate is $CuSO_4 \cdot 5H_2O$.

SOURCES OF ERROR

The following is a compilation of sources of error that students may suggest:

1. Water may still have been in the porcelain dish even after the initial heating.
2. Some water may have remained in the hydrate.
3. The anhydrous salt may have started absorbing moisture from the air while cooling.
4. Some crystals stuck to the stirring rod.
5. Some crystals spattered from the dish.

30 Measuring the Expansion of Ice

THE EXPERIMENT

PURPOSE

To measure the expansion of water as it freezes.

BACKGROUND

Water molecules are composed of two atoms of hydrogen and one atom of oxygen joined by polar covalent bonds. The molecule is bent with an angle between the two hydrogen–oxygen bonds of about 104.5°. Because oxygen is more strongly electronegative than hydrogen, the electrons are not distributed uniformly throughout the molecule but are found clustered slightly around the oxygen nucleus, forming a polar molecule. This polarity causes the water molecules to be attracted to each other; the hydrogen of one molecule is attracted to the oxygen in a neighboring molecule, and hydrogen bonds result.

When most liquids freeze, the molecules in the substance come closer together, forming a rigid, compact arrangement. For these substances, the solid is more dense than the liquid. In water, due to its molecular shape and the formation of hydrogen bonds, as the liquid freezes the molecules move slightly apart, forming a rather open, rigid hexagonal pattern. Therefore, the solid phase (ice) is less dense than the liquid phase. It is for this reason that liquid water expands as it freezes and that ice floats on water.

PROCEDURE

NOTE: Consult safety information and obtain teacher approval before beginning the experiment. Design an experiment to measure the expansion of water as it freezes.

SAFETY INFORMATION

1. Do not use any type of glass container for this experiment, including beakers, flasks, or graduated cylinders.
2. Do not fill and freeze a closed, rigid container such as a bottle.

The expansion of the ice will cause the container to burst. Be sure to allow room for the expansion of the ice.

3. After reviewing your procedure, the instructor will discuss any safety precautions that are specific to your experiment.

QUESTIONS FOR FURTHER THOUGHT

1. At what temperature is water the most dense?
2. The rapid expansion of water as it freezes can produce a lot of damage in the winter. What are some of the examples of damage that can result?
3. What are some beneficial aspects of water's expansion during freezing?

30 Measuring the Expansion of Ice

TEACHER'S NOTES

In this experiment, students are asked to measure the degree to which water expands as it freezes. This lab can be performed when students are studying the properties of solids and liquids or bonding. This lab can also be an excellent introduction to the use of constants in measurement (see Note #1 below).

OPTIONS

1. Students could compare the expansion of pure water and salt water.
2. Students could design an experiment in which they freeze water overnight and then measure the resulting volume a following day, or you can give them containers that contain a prefrozen sample of ice and have students measure the volume both before and after the ice melts.

NOTES

1. Ideally, students will calculate a constant for the expansion of ice (e.g., the percent increase in the volume); however, some students may merely state that the water expanded 2.0 mL or 5.0 mL. You can ask each team of students to place their data on the board including the initial volume of water, final volume of water, and value for the expansion. Initially the expansion values may vary quite a bit if students are merely recording the increase in volume. After discussing this with the students, they can recalculate the increase in terms of percent, and the new values should be much closer to each other.
2. The expansion of water when it freezes is approximately 9%.
3. Students should not fill and freeze the water in a sealed container or use any type of glass container. Plastic containers should be used. You may want to experiment with various containers before providing options to the students.
4. Determine how much room is available in a school freezer before deciding how many students will work together on a team.

TIME

The amount of time needed for students to design and execute their own experiment can vary depending on the level and creativity of the class, their familiarity with designing their own investigations, and the complexity of the procedure they develop.

Average Time for This Experiment

Time needed to design the experiment: 10–15 minutes
Time needed to run the experiment: 15 minutes

TEAMS

Teams of two work well. (Team size varies with freezer space.)

MATERIALS (PER TEAM)

Require the use of goggles, gloves, and aprons/lab coats. The following supplies should be readily available for the students:

- Various plastic containers
- Wax pencil
- Ruler
- Parafilm

SAMPLE PROCEDURES

1. Some students might pour a premeasured volume of water into an acceptable container, allow it to freeze, and mark the height of the ice with a wax pencil. Then the students compare the volume of the water with the volume of the ice after melting the ice and removing it from the container. To remove the ice from the container, students may place the container in a warm water bath or pour warm water over the outside of the container.
2. Some students may make graduated marks on the inside or outside of the container (assuming that it is not already graduated) before freezing the water so they can directly measure the volume of the ice.

SAFETY PRECAUTIONS

Discuss with the students any safety precautions that are specific to the procedure they develop. Material Safety Data Sheets for each substance used should be reviewed and made available to the students.

COMMON MISCONCEPTIONS, PROCEDURAL AND CALCULATION ERRORS

Students might merely give the expansion as a volume measurement. For example, if students initially started with 200 mL of water and found that the final volume of ice was 220 mL, they may conclude that the expansion of water is 20 mL rather than determining a percentage increase.

LAB REPORT

There are several alternatives to writing a formal lab report given in the introduction. If students are writing their own lab report it should include:

I. Title
II. Purpose
III. Materials
IV. Procedure
V. Data
VI. Calculations
VII. Conclusion (including a discussion of sources of error)
VIII. Answers to Questions (optional)

ANSWERS TO THE QUESTIONS

1. Water is most dense at approximately 4°C.
2. Examples of the damage caused by the expansion of ice in the winter include pipes bursting, pot hole formation, and rocks on hillsides breaking free.

3. Benefits of the expansion of water as it freezes include how ice that forms in lakes remains on the surface so people can skate or ice fish.

30 Measuring the Expansion of Ice

SAMPLE LAB REPORT

PURPOSE

To measure the expansion of ice.

MATERIALS

Safety equipment: goggles, gloves, and apron
100-mL plastic graduated cylinder
Distilled water
Parafilm

PROCEDURE

1. Pour 50 mL of distilled water into the plastic graduated cylinder.
2. Cover the graduated cylinder with parafilm to reduce the rate of evaporation.
3. Place the graduated cylinder in a freezer.
4. After the water is completely frozen, record the volume of ice in the cylinder.

DATA

Initial volume of water in the cylinder: 50.0 mL
Final volume of ice in the cylinder: 54.5 mL

CALCULATIONS

Increase in volume after freezing = 4.5 mL
% increase in the volume of the water = 4.5 mL/50.0 mL = 9.0% increase

CONCLUSION

Water expands approximately 9.0% when it freezes.

SOURCES OF ERROR

The following is a compilation of sources of error that students may suggest:

1. Even though we used distilled water, it may not have been perfectly pure because of possible contamination by the glassware.
2. The surface of the ice wasn't perfectly flat, so it was difficult to measure the final volume.
3. Despite adding the parafilm, some water may have evaporated.

31 The Popcorn Experiment

THE EXPERIMENT

PURPOSE

To determine the percentage (by mass) of water in dry popcorn kernels.

BACKGROUND

Popcorn kernels contain a certain amount of water. When the kernel is heated, the water inside becomes superheated (i.e., it remains in the liquid phase rather than vaporizing because it does not have enough room to expand). This superheated water enters the endosperm of the kernel under high pressure, causing the outercovering of the kernel to rupture. This sudden drop in pressure allows the water molecules to expand and turn to vapor. The rapid expansion of the water (and endosperm) causes the kernel to pop.[1]

PROCEDURE

NOTE: Consult safety information and obtain teacher approval before beginning the experiment. Use the following notes to develop your experiment:

1. Use between 1 and 25 kernels of dry popcorn.
2. Be sure to add a minimum of 2 tablespoons of oil to the flask before popping the kernels. (The bottom of the flask should be thoroughly covered with oil.)
3. Remove the flask from the heat source as soon as the kernels have stopped popping.

SAFETY INFORMATION

1. Do not eat the popcorn after popping.
2. Be sure to add vegetable oil to the flask before popping to avoid burning.
3. Use a one-holed stopper to prevent oil from spattering from the flask.

4. Heat the oil slowly.
5. Remove the flask from the heat as soon as popping is completed.
6. After reviewing your procedure, the instructor will discuss any safety precautions that are specific to your procedure.

QUESTIONS FOR FURTHER THOUGHT

1. If popcorn kernels are pierced with a pin before heating, the kernels do not pop. Explain.
2. When popping popcorn, there are usually some kernels that do not pop. What are some reasons for this?

REFERENCES

1. Herron, J.D. et al. *Chemistry,* D.C. Heath and Company, Lexington, MA, 1996, p. 247.

31 The Popcorn Experiment

TEACHER'S NOTES

In this experiment, students are asked to determine what percentage of a popcorn kernel's mass is water. This experiment can be used as an introduction to, or reinforcement of, percentage composition. Working with something as familiar as popcorn can provide students with a good understanding of the concept of percentage composition before relating it to more abstract substances like compounds.

OPTIONS

1. Students can compare different brands of popcorn.
2. Students could compare popcorn that has gone beyond its expiration date to fresh popcorn.
3. Students could compare popcorn that has been stored in a refrigerator or freezer to popcorn that has been stored in a cupboard.

NOTES

1. Students are told to use between 1 and 25 kernels. This is intended to generate a discussion among the team members regarding what size sample to use. They should realize that because they are working with steam, the change in mass may be very small, and therefore the larger the sample size, the better. This is a good point for you to make in any post-lab discussion.
2. A balance that reads to ± 0.01 g. will be needed, especially for teams using small samples.
3. The use of a Bunsen burner is not recommended.
4. Have insulated gloves available for the students.
5. The addition of the vegetable oil to the flask is crucial. In fact, if only a small amount of oil is added to the flask, the kernels may still burn. The bottom of the flask should be well coated with oil.
6. The use of the one-holed stopper will minimize the spattering of oil onto the hot plate as well as reduce the risk of oil burns.
7. Have soap and hot water available for the students to use when cleaning the flasks.
8. The terminology of this lab can be confusing. One person may use "popcorn" to refer to *kernels,* while another person may use the term to refer to *popped* corn. You may want to establish some terminology guidelines before beginning the experiment.

TIME

The amount of time needed for students to design and execute their own experiment can vary depending on the level and creativity of the class, their familiarity with designing their own investigations, and the complexity of the procedure they develop.

Average Time for This Experiment

Time needed to design the experiment: 15 minutes
Time needed to run the experiment: 30 minutes

TEAMS

Teams of two work well.

MATERIALS (PER TEAM)

Require the use of goggles, gloves, and aprons/lab coats. The following supplies should be readily available for the students:

- Popcorn
- Hot plate
- Flask (400 mL–600 mL work best, although 250 mL is adequate)
- 1-holed stopper to fit the flask
- Vegetable oil
- Tablespoon (optional)
- Balance
- Insulated gloves

SAMPLE PROCEDURES

Many students (see common procedural errors below) design an experiment in which they compare the mass of the flask, oil, and dried kernels with the mass of the flask, oil, and popped kernels. Whether they determine the mass of the flask, oil, and dried kernels together or separately will vary among the teams.

SAFETY PRECAUTIONS

NOTE: Also refer to the safety information given with the experiment. Discuss with the students any safety precautions that are specific to the procedure they develop. Material Safety Data Sheets for each substance used should be reviewed and made available to the students.

COMMON MISCONCEPTIONS, PROCEDURAL AND CALCULATION ERRORS

1. Students may not realize that the popcorn (after popping) will absorb the vegetable oil, and therefore they might initially mass the dry kernels and then expect to mass the popped kernels, with the difference in mass representing the lost water. However, they will find that the kernels increase in mass (rather than decrease as expected) due to the absorbed vegetable oil. Allowing students to make this error and then redesign their experiment is recommended, but you should have extra popcorn on hand for additional trials.
2. Some students may design an experiment in which they only pop one kernel of corn each trial, with the intention of performing several trials. They should discover that the change in mass for one kernel is so small that it may be difficult to measure. Students should be given the opportunity to make this error and readjust the procedure.
3. Students may calculate the percentage composition of the kernels by dividing the change in mass (which represents the mass of water lost) by the mass of the flask, oil, and kernels rather than just the mass of the kernels.

LAB REPORT

There are several alternatives to writing a formal lab report given in the introduction. If students are writing their own lab report it should include:

I. Title
II. Purpose
III. Materials
IV. Procedure
V. Data
VI. Calculations
VII. Conclusion (with a discussion of sources of error)
VIII. Answers to Questions (optional)

ANSWERS TO THE QUESTIONS

1. If the kernel is punctured before heating, the water in the kernel escapes through the opening, and therefore the pressure in the kernel never builds up to the point that would cause the hull to rupture.
2. Possible explanations include that the hull may have had an opening in it, allowing the steam to escape or the kernel may be old and dry.

FOR FURTHER READING

Borgford, C.L., Summerlin, L.R. *Chemical Activities.* American Chemical Society, Washington, DC, 1988, pp. 20–21.

31 The Popcorn Experiment

SAMPLE LAB REPORT

PURPOSE

To determine the percentage of water in popcorn (by mass).

MATERIALS

Safety equipment: goggles, gloves, and apron
500-mL flask
1-holed stopper
Vegetable oil
Hot plate
Balance
Popcorn kernels
Insulated gloves

PROCEDURE

1. Add 2 tablespoons of oil to flask.
2. Measure the mass of the flask, stopper, and vegetable oil.
3. Add 20 kernels of dried popcorn to the flask.
4. Determine the mass of the flask, stopper, oil, and dried popcorn.
5. Heat the stoppered flask on the hot plate until the corn pops.
6. Immediately remove the flask from the hot plate.
7. After the flask cools, determine the mass of the flask and its contents.

DATA

Mass of flask, stopper, and oil: 230.16 g
Mass of flask, stopper, oil, and dried kernels: 233.40 g
Mass of flask, stopper, oil, and popped kernels: 232.79 g

CALCULATIONS

Mass of dried kernels: 233.40 g − 230.16 g = 3.24 g
Mass of water lost from the corn: 233.40 g − 232.79 g = 0.61 g
% of water in corn: 0.61 g water/ 3.24 g dried popcorn = 0.19 × 100 = 19%

CONCLUSION

Dried popcorn is approximately 19% water, by mass.

SOURCES OF ERROR

The following is a compilation of sources of error that students may suggest:

1. Not all of the popcorn popped.
2. Some of the steam lost by the popcorn condensed on the inside of the flask and was therefore included in the final measurement.
3. Human and mechanical error.

32 Identifying an Unknown Compound

THE EXPERIMENT

PURPOSE

To determine the properties of seven compounds and design an experiment to identify an unknown substance.

BACKGROUND

In this experiment, you will be given seven compounds and several reagents to identify their properties. After determining the reactions and properties of the substances, design an experiment to determine the identity of an unknown compound (which is the same as one of the compounds studied in the first part of the experiment) by performing a maximum of three tests on the unknown substance. Use the following materials: (1) seven compounds—baking soda ($NaHCO_3$), sodium chloride ($NaCl$), washing soda (Na_2CO_3), plaster of paris ($CaSO_4$), chalk ($CaCO_3$), cornstarch, Epsom salts ($MgSO_4$); (2) Additional chemicals and supplies—vinegar, water, phenolphthalein, tincture of iodine, 2.0 M $NaOH$ solution, test tubes, stoppers for test tubes, test tube rack.

PROCEDURE

NOTE: Consult safety information and obtain teacher approval before beginning the experiment.

1. Perform the following tests on each compound:
   - A. the solubility of each compound in water,
   - B. the reaction of vinegar with each compound,
   - C. the reaction of iodine with each compound,
   - D. the reaction of phenolphthalein with a mixture of each compound and water,
   - E. the reaction of sodium hydroxide with a mixture of each compound and water.
2. Design an experimental procedure to identify an unknown compound with a maximum of three tests. (The compound

will be the same as one of the original seven substances studied in the first part of the experiment.)

3. Test the unknown compound to determine its identity.

SAFETY INFORMATION

1. Do not perform any additional tests without the consent of the instructor.
2. Only a small amount of product is needed for each test.
3. Do not touch or taste the substances.
4. Sodium hydroxide can cause painful burns and blindness. Follow all safety instructions carefully. If contact occurs, flush with water for 15 minutes and notify the instructor immediately.
5. Perform the vinegar test in test tubes that aren't stoppered.
6. After reviewing your procedure, the instructor will discuss any safety precautions that are specific to your experiment.

QUESTIONS FOR FURTHER THOUGHT

1. Compare qualitative and quantitative analysis.
2. Did this experiment use qualitative or quantitative analysis? Explain.

32 Identifying an Unknown Compound

TEACHER'S NOTES

This experiment is intended to be an introduction to qualitative analysis. In the experiment, students are asked to study the properties of seven compounds and design an experiment to identify one of the compounds with a maximum of three tests. The purpose of restricting the number of tests that can be performed is to force the students to develop a procedural scheme that would resemble a flow chart (i.e., after each test, the students narrow down the possible identity of the unknown compound).

After performing this lab and learning how to develop a flow chart, the students can continue with the other labs involving qualitative analysis.

OPTIONS

1. You can provide the students with a sample flow chart before the students begin the experiment.
2. You can provide the entire class with the same unknown and announce that the team that can identify the compound in the least number of tests wins. After the students complete the contest, you can display a flow chart based on the data from this experiment. Students should realize that the chart is a diagrammatic representation of the same process they developed.
3. You can provide more than one unknown for each student or team.

NOTES

1. Students may initially be frustrated with the limitation on the number of experiments that they can conduct. (Some students simply want to repeat the entire experiment and compare the results obtained on the unknown to the original data.) Any of the products used in the lab can be identified in a maximum of three tests.
2. Students will only need to use a small pea-sized sample for each test.
3. Students can add 5–15 mL of water for each test.
4. This experiment usually takes two lab periods. Students spend the first class analyzing the properties of the compounds, followed by the analysis of the unknown another day. (Students will need time to develop their experiment or flow chart before continuing to the unknown.)
5. Try not to have the known compounds on the lab tables when the students are identifying the unknown. Many students will merely compare the look of the unknown to the original substances.
6. The phenolphthalein, tincture of iodine, sodium hydroxide solution, and vinegar can all be provided to the students in dropper bottles.
7. To distribute the unknown, you can have beakers identified as A–M on your lab bench, each containing one of the seven compounds. Students can bring a small beaker or weighing paper to receive a sample of their assigned unknown.
8. When writing the lab report, the section on the unknowns should include the letter of the assigned unknown, the tests performed, the test results, and the identity of the unknown substance.

9. Students should stopper and shake the test tubes when determining the solubility of the compound.

10. When adding vinegar to the compounds, students should immediately look for a reaction. If they wait until they have added the vinegar to all of the products, they will miss the formation of bubbles. NOTE: This test should be performed in test tubes that aren't stoppered.

TIME

The amount of time needed for students to design and execute their own experiment can vary depending on the level and creativity of the class, their familiarity with designing their own investigations, and the complexity of the procedure they develop.

Average Time for This Experiment

Time needed to analyze the seven compounds: 40–50 minutes
Time needed to run the experiment on the unknown(s): 15–40 minutes (depending on the number of unknowns tested).

TEAMS

The experiment can be conducted individually or by teams of two.

MATERIALS (PER TEAM)

Require the use of gogges, gloves, and aprons/lab coats. The following supplies should be readily available for the students:

Seven compounds:
- Baking soda ($NaHCO_3$)
- Sodium chloride ($NaCl$)
- Washing soda (Na_2CO_3)
- Plaster of paris ($CaSO_4$)
- Chalk ($CaCO_3$)
- Cornstarch
- Epsom salts ($MgSO_4$)

Additional chemicals and supplies:
- Vinegar (household vinegar works fine)
- Phenolphthalein
- Tincture of iodine
- 2.0 M sodium hydroxide solution (80.0 g per liter of solution)
- Test tubes
- Stoppers for test tubes
- Test tube rack

SAMPLE PROCEDURES

(See Sample Lab Report.)

SAFETY PRECAUTIONS

NOTE: Also refer to the safety information given with the experiment. Discuss with the students any safety precautions that are specific to the procedure they develop. Material Safety Data Sheets for each substance used should be reviewed and made available to the students.

Sodium hydroxide is caustic and can cause painful burns. Students should take appropriate precautions when handling this solution. Epsom salts are toxic if ingested.

COMMON MISCONCEPTIONS, PROCEDURAL AND CALCULATION ERRORS

1. Some students have difficulty determining if the compound is soluble. If the students use a scoop of compound and only a few milliliters of water, some of a soluble compound will remain because the students have established a saturated solution. Students may mistakenly view this as an insoluble compound.
2. Students may have difficulty determining a procedure where they can identify a compound within three tests.
3. Students who don't clean the test tubes thoroughly after each test can get false results.

LAB REPORT

There are several alternatives to writing a formal lab report given in the introduction. If students are writing their own lab report it should include:

I. Title
II. Purpose
III. Data: Part 1 (on the seven known compounds)
IV. Procedure to Identify an Unknown (this may be a flow chart or written explanation)
V. Data: Part 3 (a summary of the tests performed on the unknown and their results)
VI. Conclusion (identifying the unknowns and a discussion of sources of error)
VII. Answers to Questions (optional)

ANSWERS TO THE QUESTIONS

1. Quantitative analysis involves the determination of the quantity (or amount) of substance present. Qualitative analysis involves the determination of the identity of a substance.
2. This experiment used qualitative analysis.

Acknowledgment Adapted with permission from the *Journal of Chemical Education,* vol. 68, no. 4, 1991, pp. 328–329, copyright 1991, Division of Chemical Education, Inc.

32 Identifying an Unknown Compound

SAMPLE LAB REPORT

PURPOSE

To determine the properties of seven compounds and design an experiment to identify an unknown substance.

DATA: PART 1

COMPOUND	SOLUBLE IN WATER	REACTION WITH VINEGAR	REACTION WITH IODINE TINCTURE	REACTION WITH NaOH	REACTION WITH PHENOLPHTHALEIN
Baking soda	Yes	Fizzed	Yellow*	NR	Pale pink
Washing soda	Yes	Fizzed	Yellow*	NR	Bright pink
Plaster of paris	No	NR	Yellow*	NR	NR
Chalk	No	Fizzed	Yellow*	NR	Pink
Table salt	Yes	NR	Yellow*	NR	NR
Cornstarch	No	NR	Black	NR	NR
Epsom salt	Yes	NR	Yellow*	Turned cloudy	NR

NR, no reaction.
*Students typically record these combinations as the formation of a yellow product, when it is actually not a reaction. The yellow color is the due to the iodine solution.

PROCEDURE

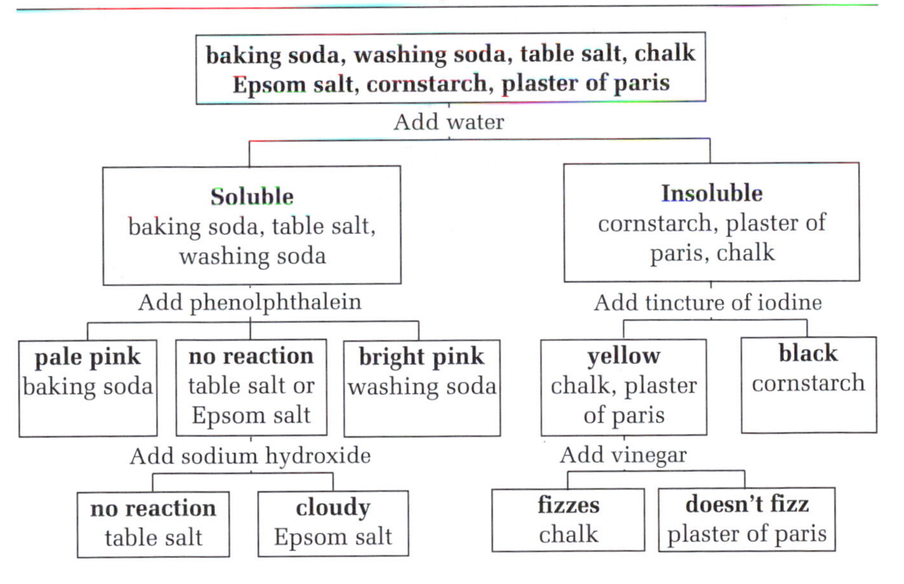

DATA: PART 3

Unknown A

Added water: it was soluble
Added phenolphthalein: it turned bright pink
Unknown A is washing soda

Unknown D

Added water: it was insoluble
Added iodine: it turned black
Unknown D is cornstarch

CONCLUSION

Unknown A is washing soda; unknown D is cornstarch.

SOURCES OF ERROR

The following is a compilation of sources of error that students may suggest:

1. Contamination of the compounds.
2. Dirty glassware.
3. Human error.

33 Using Qualitative Analysis to Determine the pH of an Unknown Solution

THE EXPERIMENT

PURPOSE

To determine the color of several indicators over a range of pH and design an experiment to determine the pH of an unknown solution.

BACKGROUND

Indicators are substances that change color as the concentration of H^+ changes (i.e., they change color with corresponding changes in pH).

In this experiment, you will be given solutions of known pH and several indicators. After determining the color of the indicators at the various values of pH, design an experiment to determine the pH of an unknown solution by performing a maximum of four tests.

PROCEDURE

NOTE: Consult safety information and obtain teacher approval before beginnning the experiment. Determine the color of the indicators at various values of pH and then design an experiment to determine the pH of an unknown solution by performing a maximum of four tests.

SAFETY INFORMATION

1. Acids and bases are both strong irritants. Contact should be avoided. If contact occurs, wash the area for 15 minutes and notify the instructor.
2. After reviewing your procedure, the instructor will discuss any safety precautions that are specific to your experiment.

QUESTIONS FOR FURTHER THOUGHT

1. How is the pH of a solution calculated?
2. What is the relationship between the pH and the pOH of a solution?
3. In a given solution, the $[H^+] = 0.01$. What is the pH and the pOH of this solution?

33 Using Qualitative Analysis to Determine the pH of an Unknown Solution

TEACHER'S NOTES

This lab can be performed after students have completed experiment 32, Identifying an unknown compound, which introduces the concepts of flow charts and qualitative analysis, or this lab can be used as introduction. This experiment can be completed when students are studying acids and bases.

OPTIONS

The number of pHs and indicators used in this experiment are left to your discretion. The greater the number of solutions involved, the more time is needed. If solutions of pH 1, 4, 7, 10, and 14 are used with 10 indicators, the first part of the lab can be completed in approximately 45 minutes if students are working in teams of two.

NOTES

1. Refer to Appendix 4 when choosing the indicators to use in this experiment. If a solution with a pH of 1 will be used, an indicator such as crystal violet or thymol blue should be available. (NOTE: Crystal violet can stain the plastic used in some types of spot plates.) Additionally, if a solution with a pH of 14 will be used, an indicator such as alizarin yellow R needs to be available. If these indicators are not available, pHs of 1 and 14 can be used in Part A, but the students will not be able to distinguish them as unknowns.
2. Buffered solutions work best for this lab. The buffered pH capsules or powders sold by many chemical supply companies can be used to prepare the solutions.
3. Using a spot plate for the reactions can reduce the volume of solutions and indicators needed for the lab.
4. Mention to the students that they should only use one or two drops of indicator for each test or the supply of indicators can be depleted very quickly.
5. The solutions and indicators can be provided in dropper bottles.
6. The same pHs used in Part A should also be used as the unknowns.
7. Students can receive the unknown in a small beaker or test tube. Only 3–4 mL of unknown are required for each student.
8. Select indicators that change color at different pHs so students will be able to identify their unknowns.

TIME

The amount of time needed for students to design and execute their own experiment can vary depending on the level and creativity of the class, their familiarity with designing their own investigations and the complexity of the procedure they develop. Times listed below will vary with the number solutions and indicators used.

Average Time for This Experiment

Time needed for Part A: 45 minutes (for 5 pHs and 10 indicators)
Time needed to identify the unknown: 30 minutes (for four unknowns)

TEAMS

Teams of two work well.

MATERIALS (PER TEAM)

Require the use of goggles, gloves, and aprons/lab coats. The following supplies should be readily available for the students:

Indicators:
- Phenolphthalein
- Bromocresol purple
- Congo red
- Bromthymol blue
- Methyl red
- Alizarin yellow R
- Methyl orange
- Thymolphthalein
- Thymol blue
- Bromcresol green

Solutions with known pH (e.g., 1, 4, 7, 10, 14)
Spot plate or test tubes
Graduated cylinders

SAMPLE PROCEDURES

Students will typically combine each of the solutions with the indicators and then design an experiment to determine the pH of an unknown solution using a maximum of four tests. Students familiar with designing a flow chart may use one as their experimental scheme, while others will write out the series of tests to be performed.

SAFETY PRECAUTIONS

NOTE: Also refer to the safety information given with the experiment. Discuss with the students any safety precautions that are specific to the procedure they develop. Material Safety Data Sheets for each substance used should be reviewed and made available to the students.

COMMON MISCONCEPTIONS, PROCEDURAL AND CALCULATION ERRORS

1. Students may think that all indicators change color at pH 7 (i.e., they are one color in acids and one color in bases).
2. Students may think that they need to use all of the indicators to identify their unknown.
3. Some students may use the entire sample of unknown for their first step and then not have any more for other tests.

LAB REPORT

There are several alternatives to writing a formal lab report given in the introduction. If students are writing their own lab report it should include:

I. Title
II. Purpose

III. Materials
IV. Data from Part A
V. Procedure to Identify an Unknown solution (this may be a flow chart or written explanation)
VI. Data from Experiment with Unknown
VII. Conclusion (including a discussion of sources of error)
VIII. Answers to Questions (optional)

ANSWERS TO THE QUESTIONS

1. The pH of a solution is calculated using the formula: $pH = -\log[H^+]$.
2. For a given solution at 25°C, the $pH + pOH = 14$.
3. If the $[H^+] = 0.01$, the pH of the solution is 2 and the pOH is 12.

33 Using Qualitative Analysis to Determine the pH of an Unknown Solution

SAMPLE LAB REPORT

PURPOSE

To determine the color of several indicators over a range of pH and design an experiment to determine the pH of an unknown solution.

MATERIALS

Safety equipment: goggles, gloves, and apron
Buffered solutions of pH 1,4,7,10 and 14
Phenolphthalein
Bromocresol purple
Congo red
Bromthymol blue
Methyl red
Alizarin yellow R
Methyl orange
Thymolphthalein
Thymol blue
Bromcresol green
Spot plate

PROCEDURE: PART A

Combine each of the solutions with the indicators and record the color changes.

DATA: PART A

	pH 1	pH 4	pH 7	pH 10	pH 14
Phenolphthalein	Clear	Clear	Clear	Pink	Pink
Bromocresol purple	Yellow	Yellow	Purple	Purple	Purple
Congo red	Black	Black	Red	Red	Red

	pH 1	pH 4	pH 7	pH 10	pH 14
Bromthymol blue	Yellow	Yellow	Blue	Blue	Blue
Methyl red	Pink	Pink	Yellow	Yellow	Yellow
Alizarin yellow R	Yellow	Yellow	Yellow	Orange	Red
Methyl orange	Red	Orange	Orange	Orange	Orange
Thymolphthalein	Clear	Clear	Clear	Blue	Blue
Thymol blue	Pink	Yellow	Yellow	Blue	Blue
Bromcresol green	Yellow	Green	Blue	Blue	Blue

PROCEDURE: PART B

To determine the identity of the unknown solutions, use the following flow chart:

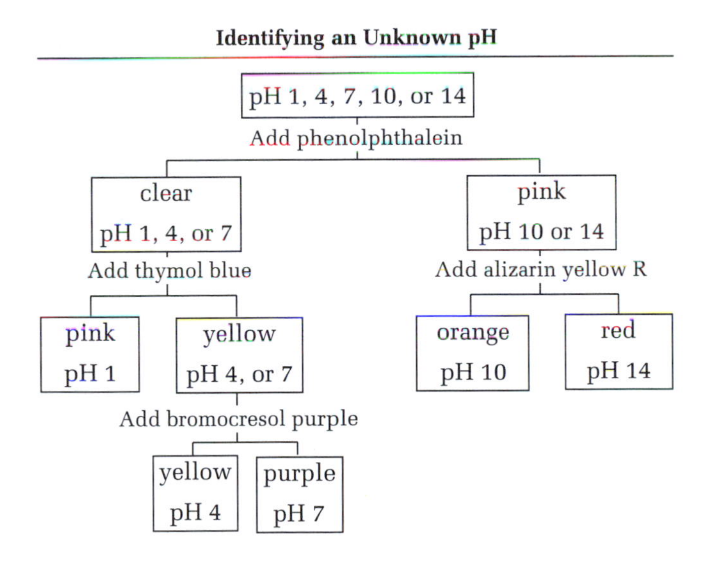

DATA: PART B

Unknown A

Added phenolphthalein: it stayed clear
Added thymol blue: it turned yellow
Added bromocresol purple: it turned purple
Unknown A has a pH of 7.

Unknown F

Added phenolphthalein: it turned pink
Added alizarin yellow R: it turned red
Unknown F has a pH of 14.

Unknown S

Added phenolphthalein: it stayed clear
Added thymol blue: it turned pink
Unknown S has a pH of 1.

CONCLUSION

Unknown A has a pH of 7; unknown F has a pH of 14; unknown S has a pH of 1.

SOURCES OF ERROR

The following is a compilation of sources of error that students may suggest:

1. The unknown solutions could have been contaminated.
2. The spot plate could have been contaminated.

34 Qualitative Analysis of the Halide Ions

THE EXPERIMENT

PURPOSE

To analyze the properties of the halide ions and design an experiment to identify an unknown halide ion in solution.

BACKGROUND

In this experiment, you will be given four solutions, each containing one of the following halide ions: F^-, Cl^-, Br^-, and I^-. Determine the properties of the four halide ions by combining them with various solutions and design an experiment to determine the identity of an unknown halide ion. Use the following materials: solutions of each of the four halide ions (F^-, Cl^-, Br^-, and I^-), commerial bleach, starch solution, calcium nitrate solution, sodium thiosulfate solution, silver nitrate solution, ammonia solution, test tubes (6), stoppers to fit test tube, test tube rack, 10-mL graduated cylinder.

PROCEDURE: PART A

NOTE: Consult safety information and obtain teacher approval before beginning the experiment.

1. Combine 5 mL of each halide solution with 1 dropperful of calcium nitrate solution.
2. Combine 2 mL of each halide solution with 5 mL of starch solution and 1 or 2 drops of commercial bleach.
3. Combine 3 mL of each halide solution with 10 drops of silver nitrate solution, followed by 5 mL of ammonia solution to each test tube that formed a precipitate with the silver nitrate.
4. Combine 3 mL of each halide solution with 10 drops of silver nitrate solution, followed by 5 mL of sodium thiosulfate solution to each test tube that formed a precipitate with silver nitrate.

PROCEDURE: PART B

NOTE: Consult safety information and obtain teacher approval before beginnning the experiment. Design an experiment to identify an unknown halide ion in three tests or less.

SAFETY INFORMATION

1. Do not combine bleach and ammonia.
2. Rinse each solution down the sink separately, with large quantities of water.
3. Silver nitrate solution is toxic and can stain the skin. If contact occurs, rinse the area with large amounts of water for 15 minutes and notify the instructor.
4. Rinse all test tubes with distilled water. Contaminants in tap water can produce false results.
5. After reviewing your procedure, the instructor will discuss any safety precautions that are specific to your experiment.

QUESTIONS FOR FURTHER THOUGHT

1. Flow charts are a valuable tool in identifying substances. If you were the president of a company that performed water testing, what economical advantages would there be to using flow charts?
2. Identify the halogens being described below:
   A. large amounts of this halogen are used in water purification
   B. the most reactive halogen
   C. a solid under standard conditions that sublimes easily, producing a purple vapor
   D. the only liquid nonmetallic element under standard conditions
   E. in solution with alcohol, this halogen is used as an antiseptic
   F. a pale greenish-yellow gas
   G. added to common table salt to aid in the prevention of goiter
3. Which of the following statements are true concerning the halogens?
   A. The halogens are nonmetals.
   B. The halogens form diatomic molecules in their elemental state.

C. The term "halogen" means salt-former.
D. The halogens are the most reactive nonmetals.
E. The halogens have a high attraction for electrons.

34 Qualitative Analysis of the Halide Ions

TEACHER'S NOTES

This lab can be performed after students have completed one of the other labs on qualitative analysis or can be used as an introduction to the technique. The experiment can also be completed while students are studying the periodic table.

OPTIONS

1. If this is the first time that students have completed an experiment that involves the development of a flow chart, you may want to spend some time discussing the concept before students begin.
2. You could introduce the concept of flow charts by providing the entire class with the same unknown and announce that the team that can identify the unknown halide ion in the least number of tests wins.
3. You can provide more than one unknown for each student or team.

NOTES

1. This experiment can be completed in one hour if students are familiar with the concept of flow charts, or the class can spend time another day working on their unknowns.
2. If this is the first time that students have worked on a lab involving qualitative analysis and flow charts, they may initially be frustrated with being limited to a certain number of tests.
3. You may want to remove all of the known solutions from the lab tables before students start the unknowns so they can't repeat any initial tests without approval.
4. You may want to have several unknowns available to the students and identify them as A–M or 1–14. Students should record the letter or number of the unknown and determine its identity. You can provide the students with a small sample (5–10 mL) of the unknown in a beaker or test tube.
5. When working on their unknowns, you may want to remind the students that the sample is all they will have to work with and therefore they will want to divide it into smaller samples for their various tests.

TIME

The amount of time needed for students to design and execute their own experiment can vary depending on the level and creativity of the class, their familiarity with designing their own investigations, and the complexity of the procedure they develop.

Average Time for This Experiment

Time need to obtain original data on known samples: 60 minutes
Time needed to identify unknowns: 5 minutes per unknown

TEAMS

Teams of two work well.

MATERIALS (PER TEAM)

Require the use of goggles, gloves, and aprons/lab coats.
NOTE: The NaF solution listed below must be prepared with distilled water. Chloride ions in tap water can react with silver nitrate to form a silver chloride precipitate, thereby producing false results.

- 0.5 M $Ca(NO_3)_2$: 118.08 g $Ca(NO_3)_2 \cdot 4H_2O$ per liter of solution
- 0.2 M NaI: 29.98 g NaI per liter of solution
- 0.2 M NaF: 8.40 g NaF per liter of solution
- 0.2 M NaBr: 20.58 g NaBr per liter of solution
- 0.2 M NaCl: 11.69 g NaCl per liter of solution
- 4 M $NH_4OH_{(aq)}$: 267 mL conc. ammonia water per liter of solution
- 0.2 M $Na_2S_2O_3$: 49.64 g $Na_2S_2O_3 \cdot 5H_2O$ per liter of solution
- 0.1 M $AgNO_3$: 16.99 g $AgNO_3$ per liter of solution
- Commercial bleach
- 3% starch solution: 30 g of soluble starch boiled in 970 mL of water
- 10-mL graduated cylinder
- Test tubes (6–8)
- Test tube brush
- Test tube rack
- Distilled water

SAMPLE PROCEDURES

Students will obtain initial test results and then design a flow chart or experimental scheme to identify the unknown halide ion. Flow charts can vary from person to person.

SAFETY PRECAUTIONS

NOTE: Also refer to the safety information given with the experiment. Discuss with the students any safety precautions that are specific to the procedure they develop. Material Safety Data Sheets for each substance used should be reviewed and made available to the students.

Students should be careful not to mix ammonia and bleach in the sink while cleaning their test tubes—doing so can produce a poisonous gas. This can be a particular problem during the portion of the lab where students are working with the unknowns and using the two reagents for their tests. Also note the following toxicities:

- Calcium nitrate: a strong oxidizer that can explode when shocked or heated, a fire risk when in contact with organic material
- Sodium iodide: moderately toxic
- Sodium fluoride: a tissue irritant
- Sodium bromide: toxic by ingestion or inhalation
- Ammonium hydroxide: the liquid and vapor are extremely irritating; avoid breathing vapors

COMMON MISCONCEPTIONS, PROCEDURAL AND CALCULATION ERRORS

1. Some students may think that they need to use the same reagents on each side of their flow chart.

2. When students are constructing their flow chart, they may forget that they added the ammonia and sodium thiosulfate to the solution containing the silver precipitate and consequently add the ammonia or sodium thiosulfate directly to the unknown halide solution, producing false results.

3. Students may have difficulty developing a flow chart or procedure that can be used to identify their unknown in three tests or less.

LAB REPORT

There are several alternatives to writing a formal lab report given in the introduction. If students are writing their own lab report it should include:

I. Title
II. Purpose
III. Data
IV. Flow Chart
V. Unknown(s)
VI. Conclusion (including a discussion of sources of error)
VII. Answers to Questions (optional)

ANSWERS TO THE QUESTIONS

1. Using flow charts to determine the identity of substances has economic advantages in the lab because it minimizes the number of tests needed. By reducing the quantity of tests, the company can save money on reagents and perform more tests in less time.
2. A. chlorine, B. fluorine, C. iodine, D. bromine, E. iodine, F. chlorine, G. iodine.
3. All of the statements are true.

Acknowledgment Adapted with permission from an experiment in *Exercises and Experiments in Modern Chemistry,* Holt, Rinehart and Winston, 1982.

34 Qualitative Analysis of the Halide Ions

SAMPLE LAB REPORT

PURPOSE

To analyze the properties of the halide ions and design an experiment to identify an uknown halide in solution.

DATA

	CALCIUM NITRATE	STARCH AND BLEACH	SILVER NITRATE	SILVER NITRATE + AMMONIA	SILVER NITRATE + SODIUM THIOSULFATE
Fluoride	Turns slightly cloudy	No reaction	No reaction	X	X
Chloride	No change	No change	Forms a white precipitate	Turns clear again	Turns clear again
Bromide	No change	No change	Forms a pale yellow precipitate	No change	Turns clear again
Iodide	No change	No change	Forms a yellow precipitate	No change	No change

FLOW CHART

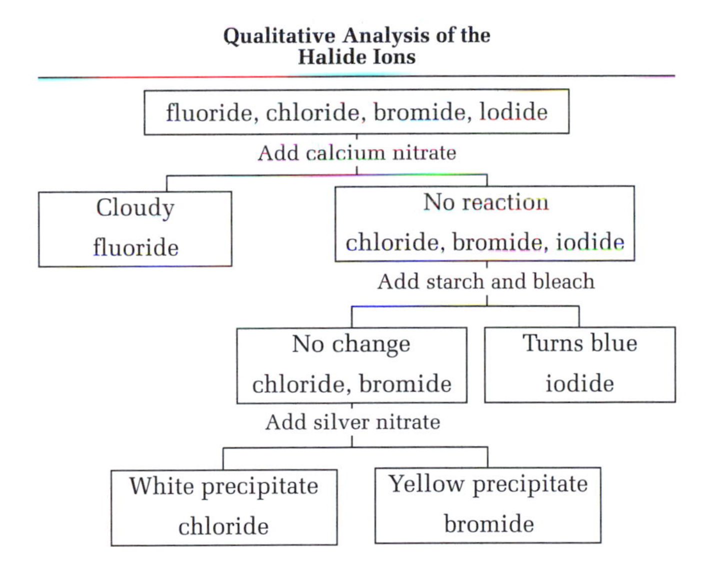

UNKNOWNS

Unknown A

Added calcium nitrate: it turned cloudy
Unknown A is fluoride.

Unknown D

Added calcium nitrate: no reaction
Added starch and bleach: no reaction

Added silver nitrate: it turned white
Unknown D is chloride.

Unknown S

Added calcium nitrate: no reaction
Added starch and bleach: it turned blue
Unknown S is iodide.

CONCLUSION

Unknown A is fluoride; unknown D is chloride; unknown S is iodide.

SOURCES OF ERROR

The following is a compilation of sources of error that students may suggest:

1. Contamination of the solutions by other classes.
2. We may not have rinsed out the test tubes well enough with the distilled water, and some contaminants from the tap water may have given us false results.
3. The test tubes may not have been cleaned thoroughly.
4. We may not have added the right amount of chemicals to each test tube.

35 Identifying Unknown Solutions

THE EXPERIMENT

PURPOSE

To determine the identity of unknown solutions by observing their reactions.

BACKGROUND

In this experiment, you will be given distilled water and a set of 7 solutions, identified only as A–H. The solutions are:

$AgNO_3$
Na_2CO_3
$Pb(NO_3)_2$
$CuSO_4 \cdot 5H_2O$
NaI
$CaCl_2 \cdot 2H_2O$
HCl

By combining A–H with each other and observing the reactions, determine the identity of each unknown.

PROCEDURE

NOTE: Consult safety information and obtain teacher approval before beginning the experiment. Design an experiment to identify eight unknowns.

SAFETY INFORMATION

1. Some of the solutions used in this experiment are poisonous. Avoid contact with the mouth.
2. Some of the solutions used in this experiment are skin and tissue irritants. If contact with the skin occurs, flush with water for 15 minutes and notify the instructor.
3. Silver nitrate and lead nitrate are highly toxic. Use appropriate precautions.
4. After reviewing your procedure, the instructor will discuss any safety precautions that are specific to your experiment.

QUESTIONS FOR FURTHER THOUGHT

1. Write balanced chemical equations for five of the reactions that occurred during this experiment.
2. During some of the reactions a precipitate was formed. Define precipitate.
3. The solutes $CuSO_4 \cdot 5H_2O$ and $CaCl_2 \cdot 2H_2O$ used in this lab are hydrates. Define hydrate.
4. Some of the combinations of solutions in this experiment resulted in a chemical reaction, while others did not. What were some signs that a chemical reaction had taken place?

35 Identifying Unknown Solutions

TEACHER'S NOTES

In this experiment, the students are given seven solutions (and water) and asked to identify each of the solutions from a list of known reagents. This lab can be used when students are studying types of reactions. Students need to be familiar with the use of a solubility table as found in Appendix 5.

OPTIONS

You can provide different sets of the same solutions, identifying the reagents with different letters.

NOTES

1. Provide the eight solutions in dropper bottles labeled A–H.
2. Students should be able to use a solubility table (see Appendix 5) to determine the formation of precipitates.
3. Students need to be familiar with the reaction of acids with carbonates.
4. It is helpful for students to use *The Handbook of Chemistry and Physics* during this lab to discover the color of certain reagents as well as precipitates. You may need to show students where to find this information in the book.
5. Some students may realize that an acid is included in the set and ask the instructor if they can use litmus paper or pH paper to test the solutions. This is left to your discretion. (NOTE: Unless the students are familiar with the hydrolysis of salts, they will be puzzled when sodium carbonate produces a basic pH.)

TIME

The amount of time needed for students to design and execute their own experiment can vary depending on the level and creativity of the class, their familiarity with designing their own investigations, and the complexity of the procedure they develop.

Average Time for This Experiment

Time needed to design the experiment: 15 minutes
Time needed to run the experiment: 30 minutes

TEAMS

Teams of two work well.

MATERIALS (PER TEAM)

Require the use of goggles, gloves, and aprons/lab coats. The following supplies should be readily available for the students:

Solutions A–H (in dropper bottles):
- Solution A: 0.1 M $AgNO_3$ (16.99 g per liter of solution)
- Solution B: 0.1 M $CuSO_4 \cdot 5H_2O$ (24.97 g per liter of solution)
- Solution C: 0.1 M NaI (14.99 g per liter of solution)
- Solution D: 0.3 M HCl (252 mL per liter of solution)
- Solution E: distilled water
- Solution F: 0.5 M $Pb(NO_3)_2$ (165.6 g per liter of solution)
- Solution G: 0.2 M Na_2CO_3 (21.20 g per liter of solution)
- Solution H: 0.2 M $CaCl_2 \cdot 2H_2O$ (29.40 g per liter of solution)

Test tubes or spot plates
pH or litmus paper

SAMPLE PROCEDURES

Students will probably mix the reagents in all combinations in test tubes or on a spot plate and then try to determine the identify of the solutions by comparing actual results with expected results.

SAFETY PRECAUTIONS

NOTE: Also refer to the safety information given with the experiment. Discuss with the students any safety precautions that are specific to the procedure they develop. Material Safety Data Sheets for each substance used should be reviewed and made available to the students. Special precautions for these reagents include:

$AgNO_3$: highly toxic and can stain the skin if contact results and the solution is not rinsed off.

$CuSO_4 \cdot 5H_2O$: body tissue irritant; toxic by ingestion. If contact occurs, wash with copious amount of water.

NaI: Deliquescent; moderately toxic. Avoid body contact. If contact occurs, wash with copious amounts of water.

HCl: Strong irritant to eyes and skin; toxic by ingestion and inhalation. Keep away from strong oxidants, alkali metals, copper, and aluminum. If contact with skin occurs, wash with copious amounts of water. If contact with eyes occurs, rinse in an eyewash for 15 minutes and call an ambulance.

$Pb(NO_3)_2$: highly toxic.

Na_2CO_3: if contact occurs, wash with copious amounts of water.

$CaCl_2 \cdot 2H_2O$: body tissue irritant. If contact occurs, wash with copious amounts of water.

LAB REPORT

There are several alternatives to writing a formal lab report given in the introduction. If students are writing their own report it should include:

I. Title
II. Purpose
III. Materials
IV. Procedure
V. Data
VI. Analysis of Data
VII. Conclusion (including a discussion of sources of error)
VIII. Answers to Questions (optional)

ANSWERS TO THE QUESTIONS

1. Chemical reactions that occurred in the experiment include:
   A. $2AgNO_{3(aq)} + Na_2CO_{3(aq)} \rightarrow Ag_2CO_{3(s)} + 2NaNO_{3(aq)}$
   B. $Pb(NO_3)_{2(aq)} + 2NaI_{(aq)} \rightarrow PbI_{2(s)} + 2NaNO_{3(aq)}$
   C. $2HCl_{(aq)} + Na_2CO_{3(aq)} \rightarrow 2NaCl_{(aq)} + H_2O_{(1)} + CO_{2(g)}$
   D. $AgNO_{3(aq)} + NaI_{(aq)} \rightarrow NaNO_{3(aq)} + AgI_{(s)}$
   E. $Pb(NO_3)_{2(aq)} + Na_2CO_{3(aq)} \rightarrow PbCO_{3(s)} + 2NaNO_{3(aq)}$
   F. $AgNO_{3(aq)} + NaCl_{(aq)} \rightarrow AgCl_{(s)} + NaNO_{3(aq)}$
2. A precipitate is an insoluble substance that forms during a chemical reaction that takes place in solution.
3. A hydrate is a crystal that contains water, chemically combined in a definite ratio.
4. Signs that a chemical change took place include the formation of a gas, the formation of a precipitate, change in color, and so on.

35 Identifying Unknown Solutions

SAMPLE LAB REPORT

PURPOSE

To determine the identity of unknown solutions by observing their reactions.

MATERIALS

Safety equipment: goggles, gloves, and apron
Solutions A–H
Spot plate

PROCEDURE

Mix each solution A–H with all of the other solutions using a spot plate. Record all observations and colors.

	A	B	C	D	E	F	G	H
A		White ppt.	Pale yellow ppt.	White ppt.	No rx.	No rx.	White ppt.	White ppt.
B			Brown ppt.	No rx.	No rx.	White ppt.	Blue ppt.	Ppt.
C				No rx.	No rx.	Yellow ppt.	No rx.	No rx.
D				No rx.	No rx.	Bubbles	No rx.	
E						No rx.	No rx.	No rx.
F							White ppt.	No rx.
G								White ppt.
H								

DATA

ANALYSIS OF DATA

Solution E is water because it didn't show any reactions.

Solution B is $CuSO_4 \cdot 5H_2O$ because it is the only blue reagent.

Solutions D and G are HCl and Na_2CO_3, respectively, because they formed carbon dioxide gas.

Solution D must be HCl because it didn't form a precipitate with copper II sulfate.

Solution G must be Na_2CO_3 because it did form a precipitate with copper II sulfate.

Solutions C and F are NaI and $Pb(NO_3)_2$, respectively, because they formed the yellow precipitate PbI_2.

Solution C must be NaI because it didn't react with solution G, Na_2CO_3.

Solution F must be $Pb(NO_3)_2$ because it formed a precipitate with Na_2CO_3.

Solutions A and H must be $AgNO_3$ and $CaCl_2$, respectively.

Solution A must be $AgNO_3$ because it formed precipitates with B,C, D, G and H.

Solution H must be $CaCl_2$ because it formed the precipitate $CaCO_3$ with solution G.

CONCLUSION

The solutions are:

A. $AgNO_3$
B. $CuSO_4 \cdot 5H_2O$
C. NaI
D. HCl
E. Distilled water
F. $Pb(NO_3)_2$
G. Na_2CO_3
H. $CaCl_2 \cdot 2H_2O$

SOURCES OF ERROR

The following is a compilation of sources of error that students may suggest:

1. It was hard to identify the formation of a precipitate in some cases.
2. The spot plate may have been dirty and contaminated some solutions.
3. The solutions may have been contaminated.
4. Some of the potential precipitates are listed as slightly soluble in the solubility table, so it is difficult to know if the precipitate will form.

Appendix 1

SAFETY RULES

GENERAL

1. Safety goggles and apron should be worn for all experiments.
2. Be familiar with, and have access to, the Material Safety Data Sheets for any chemicals used in the experiments.
3. Keep work areas neat and tidy. Leave books, backpacks, and so on in the desk area.
4. Follow all written and verbal instructions carefully.
5. Stay in your assigned lab area. Do not wander around the room or interfere with other experiments.
6. Do not eat, drink, or chew gum in the laboratory.
7. Do not touch, taste, or smell any chemicals.
8. Only perform the experiments approved by the instructor. Be sure to receive permission before beginning any experiment.
9. Know the location and proper use of all safety equipment including the first aid kit, fire extinguisher, fire blanket, eyewash, and shower.
10. Know the location of emergency exits.
11. Use a fume hood when working with poisonous vapors or volatile liquids.
12. Dispose of chemicals as directed by the instructor.
13. Check labels on reagant bottles carefully.
14. Report any accidents to the instructor.
15. Do not clean up broken glass with bare hands. Use a dustpan and broom.
16. Contact lenses should not be worn in the laboratory unless you have permission from the instructor.
17. Dress properly for the lab. Tie back long hair, roll up sleeves, remove dangling jewelry. Wear shoes that completely cover the foot.
18. Clean and wipe dry all work surfaces at the conclusion of the experiment.
19. Wash your hands before leaving the lab area.

HANDLING CHEMICALS

20. Take only as much reagant as you need.
21. Never return excess chemical to the original container. Check with the instructor regarding how to dispose of the excess.
22. When testing for an odor, use a wafting motion.
23. Do not look directly at a container that is holding chemicals.
24. When transferring reagants from one container to another, hold the containers away from your body.
25. When mixing acid and water, always add the acid to the water.
26. Do not remove chemicals from the lab area.

HEATING SUBSTANCES

27. Be sure to secure long hair and clothing around open flames and hot plates.
28. Remove all dangling jewelry.
29. Never reach over an open flame.
30. Never look into a container that is being heated.
31. Never point a test tube that is being heated toward yourself or anyone else.
32. Do not leave a lit burner unattended.
33. Do not heat anything without first obtaining permission.
34. Do not put anything into an open flame without permission.
35. Always turn off a burner or hot plate that is not in use.
36. Do not place hot glassware directly on the lab table. Always use an insulated pad.
37. Allow plenty of time for hot glass to cool before touching it.

HANDLING GLASSWARE

38. Check all glassware for cracks before beginning the experiment.
39. Never use dirty glassware.
40. Do not immerse hot glassware in cold water.
41. Lubricate glassware and glass tubing before inserting it into a stopper. Do not attempt to remove a "frozen" piece of glassware from a stopper.

Appendix 2

MASTER LIST OF MATERIALS AND CHEMICALS

Glassware

The following glassware should be available to students.

- beakers (100-mL, 250-mL, 400-mL, 600-mL)
- buret
- dropper pipette
- Erlenmeyer flasks (250 mL, 400-mL, 600-mL)
- evaporating dish
- funnel
- gas-measuring tube
- gas-collecting (or wide mouth) bottle
- glass tubing (bent and straight)
- glass plate
- graduated cylinders (10-mL, 50-mL, 100-mL)
- petri dishes
- pipette
- stirring rod
- test tubes
- watchglass

Materials

- balance
- beaker tongs
- buret clamp
- filter paper
- hot plate or Bunsen burner
- insulated gloves
- iron ring
- litmus paper
- metal cups
- mortar and pestle
- pH paper
- pinch clamp
- pneumatic trough
- ring stand
- rubber policeman
- rubber tubing
- ruler
- scissors
- spot plate
- stoppers to fit test tubes and flasks
- stopwatch
- Styrofoam cup (8- and 16-oz)
- tape
- teaspoon or tablespoon
- test tube holder
- test tube rack
- thermometers: Celsius, Fahrenheit, and blank
- thread
- tongs
- wax pencil
- weighing paper or weighing dishes
- wire gauze

CHEMICALS

- alizarin yellow R indicator
- aluminum metal (various forms)
- ammonium hydroxide solution (NH_4OH_{aq})
- antacid tablets

bleach, commercial
buffers
bromocresol green indicator
bromcresol purple indicator
bromothymol blue indicator
calcium carbonate ($CaCO_3$)
calcium chloride, anhydrous ($CaCl_2$)
calcium chloride, dihydrate ($CaCl_2 \cdot 2H_2O$)
calcium nitrate ($CaNO_3 \cdot 4H_2O$)
calcium sulfate ($CaSO_4$)
congo red indicator
copper metal (shot, wire, strips)
copper sulfate pentahydrate ($CuSO_4 \cdot 5H_2O$)
cornstarch
effervescent tablets
eggshells, brown and white
food coloring
glucose or dextrose ($C_6H_{12}O_6$)
Heat Solution™ handwarmer
hydrochloric acid (HCl)
iodine tincture
iron nails
juices (grape juice, cranberry juice)
lauric acid
lead II nitrate [$Pb(NO_3)_2$]
lead shot
magnesium sulfate ($MgSO_4$)
magnesium ribbon
marble chips
metal cylinders (copper, zinc, lead, etc.)
methyl orange indicator
methyl red indicator
paraffin candles
phenol red indicator
phenolphthalein
popcorn
potassium chloride (KCl)
potassium iodate (KIO_3)
silver nitrate ($AgNO_3$)
sodium bisulfite ($NaHSO_3$)
sodium bicarbonate ($NaHCO_3$)
sodium bromide (NaBr)
sodium carbonate (Na_2CO_3)
sodium chloride (NaCl)
sodium fluoride (NaF)
sodium hydroxide (NaOH)
sodium iodide (NaI)
sodium sulfate (Na_2SO_4)
sodium thiosulfate pentahydrate ($Na_2S_2O_3 \cdot 5H_2O$)
soluble starch
soluble ink
sucrose, table sugar ($C_{12}H_{22}O_{11}$)
sulfuric acid (concentrated)
thymol blue indicator
thymolphthalein indicator
vegetable oil
vinegar
zinc metal (1″ rectangles, mossy, strips)

Appendix 3

CROSS-REFERENCE BY TOPIC AND EXPERIMENT NUMBER

Appendix 4

RANGES OF COLOR CHANGES FOR SEVERAL INDICATORS

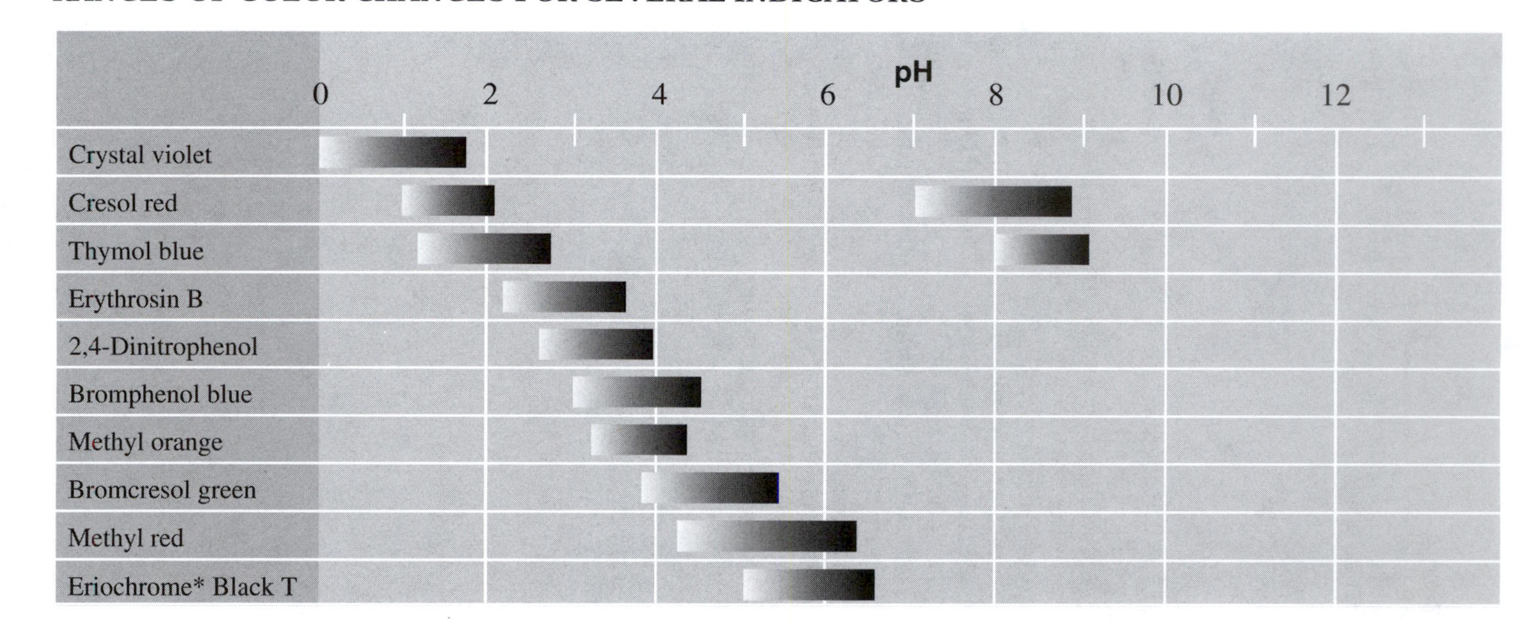

Indicator
Bromcresol purple
Alizarin
Bromthymol blue
Phenol red
m-Nitrophenol
o-Cresolphthalein
Phenolphthalein
Thymolphthalein
Alizarin yellow R
*Trademark CIBA GEIGY CORP.

The pH ranges shown are approximate. Specific transition ranges depend on the solvent.

Henry H. Holtzclaw, William R. Robinson, and Jerome D. Odom, *General Chemistry*, Tenth Edition.

Appendix 5

TABLE OF SOLUBILITIES IN WATER

	ACETATE	BROMIDE	CARBONATE	CHLORIDE	CHROMATE	HYDROXIDE	IODIDE	NITRATE	PHOSPHATE	SULFATE	SULFIDE
Aluminum	ss	s	n	s	n	i	s	s	i	s	d
Ammonium	s	s	s	s	s	s	s	s	s	s	s
Barium	s	s	i	s	i	s	s	s	i	i	d
Calcium	s	s	i	s	s	ss	s	s	i	ss	d
Copper II	s	s	i	s	i	i	n	s	i	s	i
Iron II	s	s	i	s	n	i	s	s	i	s	i
Iron III	s	s	n	s	i	i	n	s	i	ss	d
Lead	s	ss	i	ss	i	i	ss	s	i	i	i
Magnesium	s	s	i	s	s	i	s	s	i	s	d
Mercury I	ss	i	i	i	ss	n	i	s	i	ss	i
Mercury II	s	ss	i	s	ss	i	i	s	i	d	i
Potassium	s	s	s	s	s	s	s	s	s	s	s
Silver	ss	i	i	i	ss	n	i	s	i	ss	i
Sodium	s	s	s	s	s	s	s	s	s	s	s
Zinc	s	s	i	s	s	i	s	s	i	s	i

i, nearly insoluble; ss, slightly soluble; s, soluble; d, decomposes; n, not isolated. Reproduced with permission from *Chemistry: The Study of Matter,* 1987, Allyn and Bacon.

Index